Future Energy Alternatives

LONG-RANGE ENERGY PROSPECTS FOR AMERICA AND THE WORLD

by **Roy Meador**
Science Writer
Ann Arbor, Michigan

ANN ARBOR SCIENCE
PUBLISHERS INC
P.O. BOX 1425 • ANN ARBOR, MICH. 48106

Third Printing, 1979
Second Printing, 1979

Copyright © 1978 by Ann Arbor Science Publishers, Inc.
230 Collingwood, P. O. Box 1425, Ann Arbor, Michigan 48106

Library of Congress Catalog Card No. 77-92590
ISBN 0-250-40221-1

Manufactured in the United States of America

AMPLE ENERGY
How Do We Get There From Here?

Our concern in this book is with the energies of the future that man must rely on to keep his technological civilization going. A deluge of warnings in recent years has accused us of consuming traditional energies such as oil, natural gas and electricity faster than they can be replaced. How and with what they may eventually be supplemented or replaced are examined on the following pages.

When man has finished burning past deposits of energy to propel and warm the present, what can he count on in the years, decades and centuries approaching with every motion of the electric clock on the wall? Will he again adapt himself to the facts and opportunities of nature to keep his technology humming; or in helpless abandon, will he recklessly press the pedal, accelerate until the tank is empty, and then come to a sputtering stop? In this book, man's competing choices are considered in relation to the energy prospects and alternatives that contemporary technical knowledge indicates may be available.

Because we are concerned with future technology and directions, it would be imprudent to issue dogmatic certainties. There is no way to know absolutely about the future until we make the journey. Will fusion energy from the D-T cycle prove the sovereign power? Or hydrogen from water? Or that remarkable family of old reliables and versatile eccentrics, the Solar Energies? Which, if any, will be the future's first energy is information currently out of reach. But we take here the essential opening step: Learning in detail about each possibility.

As energy consumption continues at a hectic pace, fear grows that the tank may run dry before we reach the point of rendezvous with our future fuels. Fear grows that man, like extinct dinosaurs of the Upper Cretaceous, may fail to react fast enough or wisely enough. Fortunately, some men seem to have gotten the energy message. Many are now performing the basic scientific and engineering research that goes into the development of safe and reasonably economical technology to keep the lights on and the tank filled. This vital and diverse energy research is described and analyzed throughout the book.

A familiar story, perhaps already old in paleolithic times, concerns the countryman you ask for directions to a certain place. He scratches his head, dismisses several routes, and concludes ruefully, "Reckon the truth is, folks, there's just no way you *can* get there from here." As we look for an energy route to the future, there is uncertainty about the best way to go. However, we can be encouraged there are several promising energy options with conspicuous road signs pointing out accessible directions. It may take awhile, but if we check each direction, we can be optimistic about eventually getting there from here. This book seeks to provide an energy map for the coming trip.

ENERGY DEFINITIONS

From a technical dictionary: (Gr. energeia). Ability to operate or work; power to produce motion, to overcome resistance, and to effect physical changes. *binding e.,* energy equal to the difference between the weight of the nucleus of an atom and the sum of the weights of its constituent particles. *biotic e.,* the form of energy peculiar to living matter. *chemical e.,* energy which shows itself in chemical transformations. *kinetic e.,* energy in action or engaged in producing work or motion. *nuclear e.,* energy released by the splitting of the atom. *potential e., e. of position,* energy at rest or not manifested in actual work.

The Oxford English Dictionary offers this definition in connection with physics: The power of 'doing work' possessed at any instant by a body or system of bodies.

Man's future energy answers: *solar e.,* the gift of the sun. *fusion e.,* energy released by the fusing of atomic nuclei. . . *nuclear e., coal e., geothermal e., tidal e., hydrogen e.* . . . step up and take your pick.

Dedicated To

My Mother
My Sister Wanda Miller
My Brother Jerry Meador

Abundant Sources of Present and Future Energy

CONTENTS

ENERGY CRISIS OR MONEY CRISIS?

Historical proofs are lacking because it happened long ago when the human race initially began developing into the bargaining species, but a popular theory is that one of man's first complete sentences must have been this or the equivalent: "All right, you can have it, for a price." Most of us have heard countless variations on this basic offer and its popular companion philosophy that most things are available "for a price."

As a serviceman in the South Pacific, I learned our "for a price" idea is no less migratory than wild geese. It is expertly understood by rickshaw operators in Hong Kong, merchants of mementos in Manila, fruit salesmen in Formosa, and practically everyone everywhere. Today, especially, we keep colliding with the phrase in any contemporary examination of energy.

We are told that the days of cheap energy are gone, but that there is plenty of energy available "for a price." "For all we take we must pay, but the price is cruel high," says Private Mulvaney in Rudyard Kipling's classic story, "The courting of Dinah Shadd" from *Soldiers Three*. Mulvaney referred to other matters than energy, but his words aptly fit the contemporary situation. The price of energy in many places is cruel high, and it is going higher. But it is available at a price.

1

The technology capable of supplying alternate energies is ready now, and we can have energy from solar, geothermal, wind, bioconversion, nuclear, tidal and other sources for a price. The problem in connection with that fact is so simple there is no need to summon Sherlock Holmes and Dr. Watson. The crushing drawback with these energies currently is that the price is too much for individual and government pocketbooks. So what's the solution? That is equally elementary: Perfecting technology to increase efficiency and thus lower costs.

In the later years of the 1970s, the word "crisis" has been frequently used in connection with energy. A 1977 U.S. energy plan left no doubt that money and energy are inextricably united to the point where neither divorce nor surgery could break them apart. Together they spell crisis:

> The diagnosis of the U.S. energy crisis is quite simple: demand for energy is increasing, while supplies of oil and natural gas are diminishing. Unless the U.S. makes a timely adjustment before world oil becomes very scarce and very expensive in the 1980s, the nation's economic security and the American way of life will be gravely endangered. The steps the U.S. must take now are small compared to the drastic measures that will be needed if the U.S. does nothing until it is too late. How did this crisis come about? Partly it came about through lack of foresight. Americans have become accustomed to abundant, cheap energy. During the decades of the 1950s and 1970s, the real price of energy in the U.S. fell 28 percent. And from 1950 until the quadrupling of world oil prices in 1973-1974, U.S. consumption of energy increased at an annual rate of 3.5 percent. As a result of the availability of cheap energy, the U.S. developed a stock of capital goods—such as homes, cars, and factory equipment—that uses energy inefficiently.[1]

Until now we did not greatly concern ourselves with energy efficiency, because energy was cheap; and we were led to believe there was plenty more waiting at bargain prices. Then energy prices began going up, and they haven't stopped. So what has predictably developed is a crisis of *cheap* energy—with no visible symptoms of reprieve anytime soon. Thus, the energy challenge for the future is finding affordable sources and developing technologies that use less energy or harvest it economically.

CRISIS 1974 – 1977 – 19??

A successful commercial photographer in Ann Arbor, Michigan, frequently asks his clients "how much energy" they want him to invest in a particular job. He even charges for his services in terms of the biotic energy, subjectively determined, needed to complete a project.

The English poet William Blake in "The Marriage of Heaven and Hell" announced that "Energy is Eternal Delight."

Frustrated citizens in Europe, America and Japan, who waited in lines for automobile fuels during the winter and spring of 1974, might unite in paraphrasing Blake to insist that "Inadequate Energy is Accelerating Misery."

During 1974, individuals and perhaps countries were warned that energy is not an automatically available commodity, purchasable at a modest cost, and ready to go to work at the flipping of a switch. A glimpse of an energy-deficient future was provided, and the glimpse proved frightening.

A reprise of "Accelerating Misery" was not long coming. During the cold winter of 1977, Americans first felt the misery of too little natural gas and millions were asked to turn their thermostats down to a chilly 65°. Then air conditioning and other electricity needs in the hot summer of 1977 created shortages, brownouts, and for New York City a day of calamity without electricity. Shortages of energy were no longer predictions; they were realities.

The truth became starkly real that the "energy crunch" is not a vague, nebulous possibility for the 25th century, by which time science certainly will have found the perfect answer. Suddenly we were given a preview of coming distractions, and the danger was right now or just around the corner. Even during the lifetimes of current generations, all at once there was no continuing assurance of ample and constant energy.

3

In 1974 only a few weeks were needed for Americans to get back to "normal" when the fat tankers once more began guzzling their fill at the petroleum nozzles on the Persian Gulf. The grim lessons of the "No Fuel" signs were quickly forgotten by the majority of drivers, and U.S. energy consumption again ambitiously worked upward in pursuit of new records.

Then that new shocker—1977: First came the severe energy shortages of the long winter, followed by an April 18th speech on energy in which President Jimmy Carter warned with apocalyptic solemnity that "The energy crisis has not yet overwhelmed us, but it will if we do not act quickly. . . . We simply must balance our demand for energy with our rapidly shrinking resources. By acting now we can control our future instead of letting the future control us."[2] On July 28, 1977, an important energy event took place when the first oil reached Valdez, Alaska, through the multibillion-dollar pipeline crossing the state from the Prudhoe Bay oil fields. A few years previously, Alaskan oil had been seen as a vital energy reprieve for the oil-thirsty U.S., but even as the oil made its difficult way south through the accident-haunted line (the first oil needed 38.5 days for a normal 7.5-day journey) Alaska oil was already being written off as a supply that would be exhausted in little more than 25 years.

Alaskan oil did not substantially encourage the U.S. President concerning America's energy situation. On July 30, 1977, Mr. Carter said that fuel waste continued to increase, that the public had not responded well to appeals for conservation, and that voluntary compliance seemed inadequate. He suggested that "the absence of visibility to the impending oil shortage removes the incentive for the public to be concerned."[3]

THE FACTS OF ENERGY

For the well-being of mankind, the energy lessons of the recent past should be repeated again and again until the facts, however bitter, are swallowed. The energy realities on this planet are stark, critical and inescapable.

Estimates vary concerning the length of time fossil fuel reserves will be available as primary energy sources. Some contend only a few decades' supply of high-grade coal and petroleum remain. The most optimistic estimates cannot predict that world reserves of either fuel are likely to exceed three centuries as major energy suppliers.

What is unmistakably certain is that all traditional fuels are being used at a prodigious rate. In 1970, research geophysicist M. K. Hubbert noted this startling fact in an analysis of energy supplies: "In brief, most of the world's production and consumption of energy during its entire history has occurred during the last 20 years."[4] Specifically concerning fossil fuels, he concluded:

> It is clear that although the fossil fuels have been in use for about 800 years, and may continue to be exploited for a comparable length of time in the future, these fuels can serve as major sources of energy for a period no longer than about three centuries.[4]

Considering the whole of human history, past as well as future, three centuries is a minute and ephemeral branch on time's tree. And keeping in mind energy consumption during the recent past, realistic expectations for fossil fuels may be considerably less than three centuries. We are using in a matter of decades the energy supplies that it took thousands of centuries to deposit.

NATURE NEEDED OVER 40 MILLION YEARS TO MAKE A BARREL OF PETROLEUM

Robert W. Stock writing in the *New York Times Magazine* examined the unique itinerary of a barrel of petroleum.[5] He followed it through various geologic ages until it was surfaced in the Texas Panhandle during 1974 and delivered to a refinery. The most interesting part of the journey perhaps was the fuel's beginning during the Eocene period with mastodons and camels and tiny Eohippus horses roaming the Texas plains. When did that petroleum begin its development and ultimate trip to the modern refinery? More than 40 million years ago.

Even political Pollyannas so far have not suggested that we can cheerfully wait another 40 million years to replace the petroleum currently being consumed. Few householders in need of winter heat or industrial managers with wheels to turn will muster the requisite patience.

Obviously there have to be other answers than traditional fuels or we are increasingly going to find ourselves in the hard claws of crisis and shortage. Of the U.S., Stock wrote: "For a nation so dependent upon petroleum, the years ahead will be a perilous passage."[5] If depletions compel drastically reduced levels of energy consumption, extreme reversals will also inevitably be compelled in our lives. That's the message.

What is being done—what are we doing—to avoid these reversals? As the 1980s come closer, the most frequent answer seems to be: "Not nearly enough."

WILL FAITH IN TECHNOLOGY SUFFICE?

Pipedreams occasionally come true, but usually they have to be reckoned longshots. The number one pipedream in the last quarter of the twentieth century is that before old energies run out, new ones will be technically proved and ready. If not solar, then fusion. If not fusion, then solar. Something!

Faith in science and technology to find an answer in time is the determined and perhaps somewhat wistful dream we cling to rather than be blown from our moorings by harsh winds from a faithless reality. Warnings filter through from the outer darkness that on this issue technology may fall flat on its face or be painfully late in its deliveries. C. B. Reed, Energy Exploration Consultant, writes that "there is no cornucopia."[6] Reed's analysis of energy and resources emphasizes the necessity for stringent conservation and sees no reassuring technological breakthroughs close ahead.

Nevertheless, faith in technology holds on, while at the same time, ironically, technology is accused of large responsibility in causing the energy problem. Modern attitudes toward technology abound in paradox and contradiction, but underlying the most scathing criticism of our technological age is a blithe assumption that technology will not fail us. This conviction persists, and it does so perhaps because no other convenient, comforting or satisfactory conviction is available. Without faith in technology, men would have to pay more attention when they count their remaining barrels of oil. They would have to question with indignation the "easy energy" practices of the past that are still wastefully with us. These practices, prevailing in both energy development and use, have inspired consumption and waste, voluptuous and sybaritic in dimensions to rival the court of Emperor Nero. If technology could not be counted on unquestioningly, men would have to examine their "eat, drink and be merry with no thought for tomorrow" nonchalance concerning energy.

The cheap and easy energy practices were shown in U.S. oil fields when little more than half the oil typically was surfaced and used, because the oil remaining was too expensive to bring up at existing prices. More oil was left underground in some oil fields, permanently lost to mankind, than was removed. The cheap and easy practices are still shown in many of the automobiles we drive, the houses we live in, the offices we work in. We continue to use energy as if it were permanently cheap and easy, with confidence that technology won't let us down.

This is a tremendously dangerous gamble, of course, and hardly a smart one; yet it may just turn out that technology will save us from our own heedlessness and neglect. Oddly enough, with almost eerie good fortune, harnessable, reliable, unlimited new energy sources may prove to be available . . . to have been always available whenever man decided to look for them.

Falstaff in Shakespeare's *Henry IV, Part I,* informed the world that "the better part of valour is discretion." Today the better part of wisdom may be discretion in the use of energy until the new sources are given a fair chance to show what they can do.

> We are in a period of very rapid growth and change, probably the most abnormal in human history, and probably the greatest ecological and biological upheaval that the earth has ever seen. . . . We see that the power source to get us onto this rise has been fossil fuels. But fossil fuels are just a blip as far as time goes and can be depleted rather rapidly. We'll then have to depend on fission, fusion or solar energy to power the world.[7]

NEW SOURCES?

Among our jumble of possibilities, some are more possible than others. In varying degrees and ways, some are our "unlimiteds," our "infinites." To avoid being lost in the jumble, we must sort them out, determine which offer convenient, inexhaustible energy we can use, and which give more costly or dangerous trouble than they are worth.

When the sorting out is done, criteria considered, available energy outputs compared, environmental aspects weighed, and potentials measured, two energy sources tend to stand apart from the others in their promise and the hopes aroused by that promise. No precise timetable can be offered, but the assumption is that these curiously related energy sources will in the future provide the bulk of man's essential energies, with other sources playing supplementary roles.

They are already being taken for granted, perhaps complacently so, by scientists, politicians, editorial writers. An editorial in the *New York Times*, for instance, considering American coal reserves offhandedly stated: ". . . The reserves should take care of all American energy requirements from now until they are eventually met by using the sun or fusing the atom."

Solar Energy . . . Mankind's essential source of energy for biological survival since the beginning of life . . . a continuing and indispensable grant from the sun, the result of fusion reactions.

Fusion Energy . . . The sun's method adapted to earth's conditions.

In 1972 Professor Lawrence Lidsky, Associate Professor of Nuclear Engineering at M.I.T., wrote:

> There seems little question that eventually either fusion energy or solar energy will be called upon to deliver the enormous quantities of "environmentally gentle" power that man will need.[8]

It is now several years later, and the question may not be as little today as it was in 1972, but there is still long-term confidence in fusion and solar energies. Present knowledge concerning the methods and prospects for fusion and solar energy is discussed in the sections that follow. Subsidiary energy sources are also examined with care for their future potential.

So intense is research activity, new science may already be successfully exploring directions not included. An up-to-date account of mankind's energy realities is inevitably complicated by the development of entirely new information even as present data are compiled.

It is important to understand fusion, solar and the other energies man will hopefully find available when he consumes his last, aged in the earth, 40-million-year-old drink.

At the same time consideration will be given to energy and its environmental impact. Will a particular new energy prove brutal or benign? Will it fit into the recycle pattern the earth with limited and irreplaceable resources is beginning to impose on its inhabitants?

What are the energies of the future for a recycle society and how do we attain them? Shall we start with the most elusive, technologically the most complex, and if it succeeds, potentially the most rewarding: Fusion.

9

REFERENCES

1. The National Energy Plan, Executive Office of the President, Energy Policy and Planning, April 29, 1977, p. vii.
2. Carter, Jimmy. Energy Address, *The New York Times*, April 19, 1977, p. 24.
3. Clymer, Adam. *The New York Times*, July 31, 1977, p. 1.
4. Hubbert, M. K. "Industrial Energy Resources," *Nuclear Power and the Public*, H. Foreman, Ed. (Minneapolis, Minnesota: University of Minnesota Press, 1970), pp. 179-206.
5. Stock, R. W. "One Barrel of Oil," *The New York Times Magazine*, April 21, 1974, p. 14.
6. Reed, C. B. "Energy and Resources: There is No Cornucopia," *Perspectives on the Energy Crisis,* Volume 2, H. Gordon and R. Meador, Eds. (Ann Arbor, Michigan: Ann Arbor Science Publishers, 1977).
7. Gouch, W. C. "Fusion Energy and the Future," *The Chemistry of Fusion Technology*, D. M. Gruen, Ed. (New York: Plenum Publishing Corporation, 1972).
8. Lidsky, L. M. "The Quest for Fusion Power," *Technology Review*, January 1972, pp. 10-21.

CHAPTER 2

FUSION ENERGY

"Once when I asked Professor Einstein how he had arrived at his theory of relativity, he replied that he had discovered it because he was so firmly convinced of the harmony of the universe."

Hans Reichenbach

"WHEN" IS STILL THE QUESTION

In 1974 I introduced a discussion on fusion with a series of speculative questions. Three years later, those questions are still equally relevant.

Can we emulate the energy chemistries of the stars? Can we put a saddle on the energy process of the H-bomb and ride it without being thrown? Can we control and confine a gaseous plasma under high pressure and successfully discourage its wanton instabilities? Can we persuade the plasma to remain confined and to resist the innate tendency to expand? Can we achieve this obedience for what seem immense periods of time . . . several seconds? Can we regularly deliver ignition temperatures of 50,000,000° Centigrade (2.5 times the temperature of the sun) or even higher? Can we devise a container capable of withstanding pressures, temperatures and neutron bombardments, while maintaining a hospitable environment for its unstable "star boarder," the critical plasma? Can we develop a substantial family of new technologies, some we know about, some we can guess, and some no doubt lurking slyly on this side of the "scientific feasibility" horizon? Finally, if these questions are satisfactorily resolved, can we efficiently recover any energy that results and put it to work for man?

The fact that none of these questions can be definitively answered after three years of research effort in physics laboratories around the world is no reason for discouragement. Impatience, of course, is a familiar human characteristic. If something extraordinarily promising and attractive such as fusion energy is offered, we want it now, or at least soon. "When?" we want to know. "This year, next year, the year after?"

Scientists and nuclear engineers pay little attention to such impatient queries. They go on about their business, seeking to prove the practicability of electricity from fusion in a safe, controlled, nonpolluting process that will supply the energy needs of mankind potentially for thousands of years. The prospect of fusion energy is so splendid to contemplate, we feel impatience growing stronger. Why wait? Why not throw ourselves into a mammoth worldwide research effort, spend what it takes, and have fusion energy sooner rather than later?

Additional funds intelligently invested at fusion research centers might hasten the day when the difficult questions surrounding this future energy are answered. But fusion still continues to be largely a matter, at this stage, of basic science. First the scientific building blocks of fusion must be put in place. Then perhaps the worldwide effort to construct an energy source on those blocks will be easier to promote. Fusion might be likened now to atomic research in the 1930s. During those years, research was carried on quietly and privately by many workers in many countries. In 1938 nuclear fission was demonstrated by Otto Hahn and Fritz Strassmann. The meaning of the Hahn-Strassman findings was supplied the following year by Lise Meitner and Otto Frisch. Experiments were widely conducted, fact was added to fact. Then Enrico Fermi and a research group achieved the first sustained nuclear reaction at the University of Chicago in 1942. The successful activation of an atomic pile in Chicago has been called by some the beginning of the Atomic Age. Today a sculpture by Henry Moore stands on the site of the initial atomic reaction to commemorate the event. Appropriately, the statue is called simply "Energy." A case could be made, however, that the Atomic Age began earlier

12

with Rutherford's 1911 theory of atomic structure, or Niels Bohr's application of the Planck-Einstein quantum theory to orbital energy, or Chadwick's 1932 discovery of the neutron, or the splitting of lithium atoms with accelerated protons by Cockcroft and Walton the same year, or Fermi's 1933 achievement of nuclear changes with slow neutrons. All these efforts collectively added up to reach the point where an all-out scientific-engineering struggle, the Manhattan Project in the United States, would work.

Fusion has not yet reached the point of successful demonstration. So it is premature to say that the Fusion Age has begun. But the experiments and step-by-step findings that will lead to that beginning are taking place steadily. In that sense, the Fusion Age is well underway. Sometime in the future, science historians will look back to these years and the current fusion experiments taking place, as the start, just as we look back to the early decades of the twentieth century for the birth pangs and development stages of atomic energy.

However splendid fusion energy sounds, what we need for the time being is patience. The difficult technical challenges of fusion are not going to be solved in a few years. Scientists speak in terms of decades when they speculate about the fusion timetable. To them, 30, 40 or 50 years doesn't seem too long to wait for an energy source that will provide clean energy, with none or few radioactive wastes, and that has the "potential of minimum environmental insult."[1]

Instead of asking "when" impatiently, we should now encourage fusion scientists with a friendly, "Keep at it!"

THE 4.5-BILLION-YEAR ENERGY SOURCE

Approximately 4.5 billion years ago the earth was created, and its source of power throughout this brief existence has been fusion energy from the sun. The sun provides awesome proof of the energies available through fusion. Fusion energy from the sun is produced by the thermonuclear fusing of two protons or hydrogen nuclei ($_1H^1$) to form a nucleus of deuterium ($_1H^2$), which fuses with another hydrogen nucleus to form an isotope of helium ($_2$He^3), which in turn fuses with another helium isotope to form an ordinary helium nucleus and two protons. What is the by-product released by this continuing process? Tremendous ENERGY. And estimates are that the same fusion process, which has been delivering energy to the earth constantly since the earth's creation, can feed on the mighty reservoir of solar hydrogen for possibly another 30 billion years.

Is that all we have to do, imitate the stars, in order to achieve this bounteous legacy of fusion energy? Then what are we waiting for, why not bring hydrogen and helium together, allow them to play around in the fashion of the sun, and stand ready to catch the energy? Well, unfortunately, it isn't that easy. There are problems: Finding practical answers to those plaguing questions with which we started.

THE FUSION QUEST IS ON

The sun's energy secret is thought to be known, although scientific questions are sometimes raised to challenge that assumption. Whether or not fusion fully explains the energy of the sun, scientists, governments and industries are increasingly active and determined in their effort to solve the problems and make controlled fusion energy a reality for man. The steady depletion in fossil fuels, themselves deposits of fusion energy from the sun, and the accelerating demand for new, long-term energy sources, make the effort important and will undoubtedly lead to its steady intensification.

The inescapable impression is that fusion is an insistent fact looking impatiently for practical applications, and that it offers the world its chief energy hope for the long future. Both the literature and the broadening scope of research compel the conclusion that if fusion is not an idea whose time has come, it is an idea determinedly going out to find its own time. Mankind needs fusion energy as quickly as possible, and what mankind needs a few men always try to get.

14

Fusion research today is well into its second decade on a worldwide scale. One research program at the University of Michigan is close to its "crystal" or 15th anniversary of continuous work. This program, under the direction of Professor Terry Kammash, is concerned with the theoretical aspects of plasma behavior in magnetic fields, a key consideration for the ultimate effectiveness of the fusion process.

In his book, *Fusion Reactor Physics: Principles and Technology,* published in 1975 and now in its third printing,[2] Kammash writes:

> For about a quarter of a century many countries throughout the world have been engaged in a research aimed at producing power from fusion nuclear reactions. The primary motivation is the availability and easy accessibility of an inexhaustible source of fuel for use in fusion reactors. Although it only exists as one part in about 7000, there is enough deuterium in the ocean waters which if totally burned will meet mankind's energy needs for millions of years. It is for this reason that all these nations are counting on controlled fusion as the long-range answer to the energy crisis.

Deuterium (an isotope of hydrogen, also called "heavy hydrogen") is considered an inevitable component in the fusion fuel cycle. The word "inevitable" is not casually applied. Most scientists engaged in fusion research are convinced that eventually fusion energy will be available, and will become a leading contributor to mankind's energy supplies.

Science writer Isaac Asimov, after examining the pros and cons of fusion progress so far, notes that fusion could make all nations "have" instead of "have not" nations. In 1971 the Director of Long Range Planning at the Oak Ridge National Laboratory, David Rose, wrote: "The present consensus is that, scientifically speaking, controlled fusion is probably attainable."[3] In 1974 Dr R. F. Post, Lawrence Livermore Laboratory, and Dr. F. L. Ribe, Los Alamos Scientific Laboratory, wrote that fusion should be considered the ultimate energy source, with other sources developed as interim sources.[4] In 1977 public news releases continued to emphasize the continuing quest and the persistence of the fusion effort. In mid-1977 the Argonne National Laboratory of Illinois reported the design of a new

15

reactor called the Prototype Experimental Power Reactor-Ignition Test Reactor, a Tokamak type. This is the reactor, according to the designers, that will be available by the mid-1980s and will prove once and for all the feasibility of fusion.

This optimism was balanced by the less confident, but nevertheless positive statement on fusion included in the 1977 National Energy Plan published by the Executive Office of the U.S. President.

> Research in controlled thermonuclear reactions ("fusion") has been a major element in energy research and development programs. However, despite many years of active research, scientific feasibility has yet to be demonstrated, though steady progress has been made in satisfying each of the individual criteria for achievement of breakeven power (the production of more power than is consumed). Current research on magnetic confinement systems seeks to demonstrate the simultaneous attainment of temperature, density and confinement time necessary for breakeven. Inertial confinement (laser or beam) systems, a newer technology, may lag behind magnetic systems in achieving breakeven power. Once a demonstration of breakeven is made, extensive engineering efforts would be required to design a commercial system. However, even without achievement of breakeven power, either fusion system may be able to produce usable energy as part of a hybrid fusion-fission cycle. The fusion process produces neutrons which might breed fuel for light-water nuclear reactors more easily than it produces electricity.[5]

This National Energy Plan concludes with a request for a continuation of fusion research. Whether this request was made or not, research would certainly go on. At this point, it would be impossible to halt or even significantly impede the momentum of science in its pursuit of fusion answers. Too many scientists and countries are working on the problems, and the problems are much too intriguing to drop. Professor Kammash and other scientists believe that fusion research is pretty much "on schedule." They have no doubt that their mathematical demonstrations will eventually be proven experimentally. Kammash thinks the feasibility of controlled fusion reactors may be established in the 1980s, with commercial energy from fusion sources during the first half of the 21st century. Even those scientists who offer more cautious predictions persist in their conviction that fusion energy will eventually be a reality.

BACKGROUND OF THE FUSION QUEST

Over 40 years ago nuclear fusion was experimentally accomplished on earth. In 1932, an experiment showed the possibility of releasing fusion energy through collisions of accelerated deuterium nuclei, but the possibility was not taken seriously as a practical means of power production until much later. Gough and Eastlund consider the reason for this neglect the fact that in the particle accelerators used then, colliding deuterium nuclei would scatter without reacting, making it impossible to achieve greater energy than that needed to start the acceleration of nuclei.[6]

But new knowledge opened new possibilities. The first thermonuclear test explosion in 1952 proved that elevating the temperature of gaseous collections of electrically charged particles (plasma) enormously could set off fusion reactions and consequently release fusion energy.

> Coincident with the development of the hydrogen bomb, the search for a more controlled means of releasing fusion energy was begun independently in the U.S., Britain and the U.S.S.R. Essentially this search involves looking for a practical way to maintain a comparatively low-density plasma at a temperature high enough so that the output of fusion energy derived from the plasma is greater than the input of some other kind of energy supplied to the plasma. Since no solid material can exist at the temperature range required for a useful energy output (on the order of 100 million degrees C) the principal emphasis from the beginning has been on the use of magnetic fields to confine the plasma.[6]

THE CHALLENGE ESTABLISHED

Those basic technical facts and related developments set the stage for current fusion research. A controlled fusion reactor will have as its underlying principle a process involving the heating of plasma to temperatures as high as 100 million degrees C. When that occurs, if other critical requirements are met, fusion will take place through the joining of two light atomic nuclei (deuterium) to form a heavier nucleus and releasing usable energy. The identical process, with different fuel cycles, is believed to energize the stars.

> When two light atomic nuclei are brought together with enough force to overcome the repelling Coulomb force, they fuse, yielding a heavier nucleus and at least one other particle, a proton or neutron depending on the reaction.[7]

The fundamental problem right now seems to be consistent and reliable confinement of plasma. For fusion to occur, the plasma must be rigidly controlled or "pinched" into a regulated flow so that the total effect of the ignition temperature will be experienced and retained for a sufficient period to allow thermonuclear activity. The plasma must be prevented from "kinking," or breaking, or losing heat by touching the walls of its container. Such dissipation of heat can destroy the conditions essential for fusion.

Magnetic fields are used to "corral" the plasma, or to keep charged particles from reactor walls and under control. This corral for plasma is sometimes called a "Magnetic Bottle," with lines of force inside a magnetic coil parallel to the plasma column and in effect "stiffening" it for the occurrence of fusion.

Minimizing the difficulties of the challenges involved would be naïve. A suitable fuel cycle must be determined, in itself a mammoth undertaking. Then a container must be designed to enclose the fuel cycle. For a number of years, most experts have believed that magnetic confinement will probably be part of the ultimate solution. Nevertheless, full-scale studies of alternate approaches have not been neglected, and as fusion research goes forward, there will probably be few roads not taken by searching scientists. "Two roads diverged in a wood, and I—/I took the one less traveled by,/And that has made all the difference," wrote Robert Frost. Fusion researchers, sometime in the future, may look back and note a little-traveled road that made all the difference. Meanwhile, fusion research is occurring in nearly every technologically advanced country. Many roads are being investigated, many theories evaluated, many procedures tested.

In current work on the confinement problem, the main concern is still that of understanding and eliminating "instabilities." This is the term assigned by plasma researchers to the tendency of particles confined in a magnetic field to move independently in ways that defy confinement. Even the basic particles of nature seem to have freedom drives quite independent of human politics. To achieve fusion power, "obedient" plasmas are required. The container is where this obedience theoretically will be achieved.

PLASMA CONTAINERS

A preeminent necessity for success in the fusion quest will be "plasma containers" assuring magnetic or inertial confinement of plasma and providing a suitable climate or site wherein fusion can occur and resultant energy harvested. Thanks to cooperative international efforts, progress is being substantively made in the development of plasma containers.

> Recent advances in the performance of several experimental plasma containers have brought the fusion power option very close to the "break even" level of scientific feasibility.[6]

This confidence is sustained by results of experiments using Tokamak and Stellarator plasma confinement apparatus in the U.S. and the U.S.S.R.

Tokamak

The Tokamak-type plasma containers were principally developed at the I. V. Kurchatov Institute of Atomic Energy in Moscow. A Tokamak type has also been put into operation at Oak Ridge, at the Plasma Physics Laboratory of Princeton University, and at other research sites in the U.S. Tokamak is a toroidal machine confining hot plasma on an intricate surface of helical magnetic field lines. In the Tokamak, the plasma is "doughnut-shaped" and subjected to a constricting force generated by the current through it. The main constricting force, generated by an external current, is called a "toroidal pinch." A

19

toroidal diffuse pinch has the current over the cross section of the plasma.[7]

The cooperative nature of the world fusion research attitude has made the Tokamak an internationally accessible apparatus with corresponding benefits for all.

In the mid-seventies, the Tokamak was considered the most promising approach to the achievement of controlled fusion. Four large Tokamaks were being constructed (in the U.S., Russia, Japan and Italy, cooperatively for the Common Market countries). The reactor being designed in 1977 by the Argonne National Laboratory, Illinois, and the General Atomic Company, San Diego, was announced as a "Tokamak-type . . . in which the fuel isotopes are stripped of their electrons and magnetically suspended as a charged gas in a doughnut-shaped chamber."

At the same time, competing devices were still receiving both research effort and funds because doubts lingered whether or not the Tokamak road was the one leading to the energy pot of gold.

Will "Big Tokamak" Establish Feasibility?

Tokamak is the anglicized version of a Russian word meaning "magnetic twist" or "configuration." Tokamak must twist plasma magnetically into a doughnut shape and hold the shape long enough for fusion. Holding the doughnut shape has been a troublesome problem with the smaller Tokamaks. If plasma breaks from its configuration too soon, the slightest contact with the reactor walls causes pollution. Among the big Tokamaks under construction has been the TCT-TFTR (Two-Component Tokamak, Tokamak Fusion Test Reactor) at Princeton University. The TCT-TFTR if it lived up to expectations would show progress in the critical matter of plasma confinement time. The hope was that it would increase plasma confinement time from the approximately 20 milliseconds possible with smaller Tokamaks.

> More than any other single problem, impurities introduced into plasmas from tokamak walls seem to threaten the success of tokamak fusion now, because the impurities, which have higher

20

atomic numbers than the hydrogen ions, can radiate away the heat of the plasma and hold down its temperature. Physicists will want to learn how well the TCT-TFTR handles the problems of impurities, since it was intentionally designed as an oversized machine for its temperature and containment to keep the plasma away from the walls.[8]

Japan's big Tokamak was to have a design similar to the TFTR. The JET (Joint European Tokamak) was designed at Culham, Great Britain, for location at Ispra, Italy. Finally there was the Russian T20, the fourth big Tokamak. Applying lessons learned with smaller models, the goal with the big Tokamaks is to hold the doughnut shape longer and attain confinement times adequate for fusion. But as noted, all research eggs are not being assigned to this special basket.

Stellarator

This is another promising toroidal machine used in Russian and European research. Confinement of plasma is effected by field coils producing an axial magnetic field. The helical windings of the coils supply the pinch force that deters plasma from migrating to the container walls, where temperature peaks are flattened and fusion cannot take place.[9]

Other confinement approaches include the mirror machine at the Lawrence Livermore Laboratory in California. In this form of open-ended device, energy losses at the end are a problem conceivably necessitating direct energy conversion. Another valid confinement approach that may have a chance to win the race to the ever closer plateau of scientific feasibility is the Theta-Pinch device at the Los Alamos Scientific Laboratory.[2] Coppi and Rem set down general guidelines as follows:

> A fusion reactor could consist simply of a container holding a mixture of fully ionized deuterium and tritium nuclei at a very high temperature. In such a hot plasma fusion reactions would occur when the ignition temperature is reached; at this temperature the energy released by the fusion reactions equals the energy lost by radiation from the plasma. Even for the deuterium-tritium reaction, which has the lowest ignition temperature, this temperature is still quite high: 46 million degrees Kelvin. Clearly a plasma in which thermonuclear reactions will

begin cannot be contained by material walls. Most of the envisioned fusion reactors are therefore based on a plasma confined by a magnetic field.[7]

Whatever plasma container proves to be the ultimate answer, its
function can be simply stated: To provide an adequate arena
where the process of fusion can and will occur.

FUSION FUEL CYCLES

For fusion reactors, a number of fuel arrangements are under
consideration:

	Estimated Ignition Temperature[10]
Deuterium + Tritium (D-T Cycle)	50,000,000° C
Deuterium + Deuterium (D-D Cycle)	300,000,000° C
Deuterium + Helium^3 (D-He^3 Cycle)	500,000,000° C
Hydrogen + Lithium^6 (H-Li^6 Cycle)	900,000,000° C

The temperature figures are formidable, but Gough notes that
"we have exceeded in our experiments the ignition temperatures
for all four fuel cycles. We are working over a wide density
range, many tens of billions. And we can have fusion reactors
over a range of one billion in density."[10]

The cycle with the highest reaction rate and the lowest temperature for fusion is the D-T cycle. Because of these facts, the
Division of Controlled Thermonuclear Research for the Atomic
Energy Commission calls the D-T cycle the "most attractive for
first-generation fusion reactors."[1]

The D-T cycle requires deuterium and tritium. The first is
easy, the second is a problem. Tritium is derived from lithium,
a mineral currently in short supply. In 1974 worldwide production of lithium was about 5 million kilograms, with approximately 3.4 million kilograms of this total produced by the U.S.,
the world's chief producer and exporter of lithium. With
lithium also expected to be a vital ingredient in energy-efficient
lithium-sulfur batteries, success of the D-T cycle could mean
that fusion energy from the start would be haunted by potential
shortages. At present, long-range reserves of lithium are un-

certain; but the mineral has not yet received the extensive exploration effort that will follow expansion of the market.

The need for a lithium reaction to breed tritium (possibly through neutron absorption in natural lithium which produces tritium and inert helium) makes the D-T cycle technically more complex than the D-D cycle, but the advantages of reaction rate and temperatures are expected to make the D-T cycle at least initially preferable.

Lidsky notes that although the D-T reaction probably will be exploited first, we need not be overly concerned about running out of lithium. Supplies of lithium are sufficient for several hundred years, and then:

> . . . it is hard to believe that we will not have gained sufficient expertise in plasma physics during that time to enable us to tap the energy content of the D-D reaction.[11]

When technology reaches the point it is expected to reach, where a straight deuterium fuel cycle can be utilized for fusion energy, mankind's fuel anxieties should be reduced and possibly may be eliminated. Calculations indicate that fusion of the deuterium nuclei in one gallon of water could release the energy obtainable from combustion of 300 gallons of petroleum.

And here's the whipped cream on the strawberries: Deuterium as a constituent of all water, including the oceans, is easily and inexpensively available. A few cents will cover the cost of extracting all the deuterium from a gallon of water.[9] When technology develops the skills to take advantage of nature's generosity, the inexpensive abundance of deuterium could end our energy woes and allow the human race to concentrate on other problems. To reach that point, all we have to do is prove that it is possible, and then carry it out. Bring on the magicians.

THE THREE CRITERIA OF FEASIBILITY

To prove the scientific feasibility of fusion as a reliable energy source, three major challenges must be met involving: (1) density, (2) confinement time and (3) temperature.

23

The distance between what has been achieved and what must be achieved has been progressively reduced for all three in separate experiments. At M.I.T., using the Alcator (Tokamak-type), the product of density and confinement time has reached the high level of 10^{13} cm^{-3} sec.[12] For fusion to take place, the level of 10^{14} cm^{-3} sec must be achieved. These encouraging results were reached without a technical breakthrough, but simply by applying known technologies to produce a more refined plasma.

Using the 2X-IIB mirror machine at Lawrence Livermore Laboratory, temperatures exceeding 100 million degrees have been attained. These are believed accurate burn temperatures for fusion reactions. An ERDA–CTR report indicated that a temperature of 13 keV was accomplished, and noted that the experiments "incorporated a warm plasma stream injected along the magnetic field lines which suppressed the ion cyclotron instabilities . . . . Neutral beam injection into the streaming target plasma increased the mean ion energy from 3 keV to 13 keV."[13] Temperatures reached by the 2X-IIB strengthen convictions that feasibility can be demonstrated in the 1980s.

A neutral beam heating technique is used in the 2X-IIB. Energetic, heavily charged particles, the same materials as the plasma, are fired into the plasma gas at high velocity in simultaneous volleys from high-power injectors (analogous to a water spray with many nozzles). As particles collide, the plasma is heated. The same approach is being tried with variations in other research.

Thus, the feasibility point is brought closer through the use of a purer plasma in the Alcator to improve density-confinement times while, separately, neutral beam heating supplies the temperatures required for fusion. But the task is a great one, and for every small advance there is a mammoth journey still ahead. Scientific workers continue striving to make improvements. They know now that greater plasma purity is achieved by scrubbing the inner walls of Tokamak chambers to prevent contamination. Different metals, such as aluminum and various alloys, are also being considered for Tokamak walls to prevent contamination and extend density-confinement time.

LASER FUSION–AN ALTERNATIVE POSSIBILITY

Speaking at Stanford University in 1974, Dr. Edward Teller said of laser fusion: "It won't help us with this energy crisis or the next one. Maybe it will help us with the one after that."[14]

Thermonuclear energy in small amounts has been released through laser bombardment of a compacted deuterium and tritium pellet. The pellets were tiny glass spheres, called microballoons, containing deuterium and tritium in gaseous form. The walls were approximately 1 micrometer thick. A neodymium glass laser concentrated two laser beams on the sphere by means of elliptical mirrors. The lasers heated the glass wall, driving it inside to compress the gaseous fuel. The temperature at the core was raised approximately to 10 million degrees. Neutrons were emitted in what was called a fusion reaction, although some critics have challenged this interpretation, contending that the temperature in this experiment is not high enough for a true "thermonuclear burn."

This laser beam experiment was conducted by KMS Fusion, Inc., Ann Arbor, Michigan, and has been successfully repeated by others. That such experiments have achieved neutrons is not disputed. This means that fusion has taken place, whether in the gaseous core or in a few isolated, fast-moving ions.[15] The KMS results do not prove, but do support the basic theory of laser fusion that concentrating a fast laser on an extremely minute spot will result in compression, very high temperatures, and perhaps . . . fusion.

Because of these findings, a thermonuclear burn producing billions of neutrons and an incontestable fusion reaction is hoped for when Lawrence Livermore Laboratory's large laser system, SHIVA, is fully operative and experiments are carried out on a major scale.

FUSION NOW WITH NO WAITING

Serious proposals have been advanced concerning a way to obtain fusion energy immediately and explosively. The method

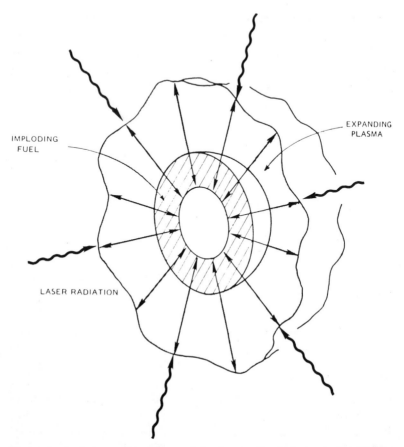

IMPLODING
FUEL

EXPANDING
PLASMA

LASER RADIATION

Figure 1. Artist's representation of the laser-pellet fusion reaction. From: *Fusion Reactor Physics: Principles and Technology* by Terry Kammash, Ann Arbor Science Publishers, Inc., 1975.

is simple. A hydrogen bomb would be employed as a fusion re-actor. Set off underground, the bomb would release vast quan-tities of thermonuclear energy which could be captured for use. Salt dome cavities along the Gulf coast of the United States have been suggested as possible sites. The same cavities have also been recommended as storage places for oil reserves, giving us a curious example of competition among the energies.

As proposed, two thermonuclear detonations might take place daily at an energy plant of this sort. The explosions would deliver superheated steam which would operate electrical generating apparatus and at the same time provide nuclear fuel for fast-breeding reactors. This is another energy frontier waiting to be explored, though some might think of it as an energy wilderness. Advocates contend fusion energy could be available in 10 years rather than 50 using this approach. Would the walls of the cavity withstand the explosions and the high temperatures? These are questions to be answered very carefully before rushing into this technique of thermonuclear fusion, but it is reported that "the seismic effect of an underground nuclear blast is reduced 100-fold if the blast is conducted in a large cavity."[16]

An estimated seven hundred 50-kiloton explosions would be needed annually. With the cost of one 50-kiloton thermonuclear device approximately $420,000, the procedure would not be economically practical until this figure had been reduced by nine-tenths. Supporters, despite the science fiction rather than science fact dimensions of their proposition, want a chance to prove that both costs and explosions can be kept within acceptable limits. Man's need for more and more energy may eventually provide the chance.

SCIENTIFIC ADVENTURE IN PURSUIT
OF A HAPPY ENDING

This strange but quite serious energy alternative indicates rather dramatically that no fusion stone is being left unturned. Despite setbacks, fusion research in all directions will no doubt continue internationally, with scientists around the world cooperating in one of the greatest scientific adventures of all time. Breakthroughs in science occur when scientists keep busy, in many different places exploring many different avenues.

Put 10,000 monkeys to work at typewriters, we're told, and they'll eventually turn out the plays of Shakespeare.

Much more likely is the surety that if 10,000 scientists are focused on a challenge whose solution is both scientifically possible and pragmatically useful, the problem will be solved. Clearly, problems as complex and manifold as those involved in fusion are going to be solved one at a time by diverse groups and individuals working in different parts of the world. Success in the fusion quest will undoubtedly prove in the end to be international in scope.

A relevant analogy might be the impressive results achieved in the scientific exploration and understanding of Antarctica since the International Geophysical Year (1957-58). Today, a vast continent, covered with ice and substantially unknown by man for 11 million years, is the scene of remarkable international collaboration. Nationals who might quarrel over political sorespots at home cooperate in the Antarctic, because cooperation is the only way to succeed.

It doesn't seem farfetched or inaccurate to call fusion today's "Antarctica of Science." In a real sense fusion is our most important undeveloped continent, and it genuinely seems as if we are now on the threshold of knowledge, as well as understanding and conquering the mysteries there. During the IGY, scientific task forces from many nations appeared on the icy doorsteps of the Antarctic and began digging in. Fusion has not yet had an International Fusion Year, or what might be more realistic, an International Fusion Decade. However, the scientists of many countries are "digging in," and the results will be to unmask fusion as comparable effort is so effectively unmasking the Antarctic.

Failures in experiments, disappointments and inadequate funds may intermittently slow but won't stop the fusion advance. The international scientific effort will go on because among all the current energy alternatives fusion continues to be man's best hope for striking it rich. The incurable old prospector impulse in the human race will keep us striving for fusion as long as there is a mustard seed of hope for success.

And the hopes for fusion are by no means dimmed because the pace of progress is slow. Even practical businessmen are beginning tentatively to count their fusion eggs. Potential major users such as electric utility companies are actively interested. Clinton P. Ashworth, spokesman for a West Coast utility, expressed this conviction:

It is possible that a fusion reactor anyone in this country is going to want to buy is yet to be invented. However, it seems more probable that pieces and ideas for a practical fusion reactor exist but have yet to be assembled into self consistent reactor concepts and their technical gaps closed . . . . I do believe that we will need a real fusion energy option and the sooner we can get it the better.[17]

The very complexity of the technological challenges thrown at us by fusion as well as the long delays between steps conspire to protect us from the warnings issued by V. L. Parsegian about investing total faith in a single energy.

It would be unfortunate, however, if "fusion enthusiasm" were to detract from substantially larger funding for the development of improved fission reactors . . . . A carefully planned program for achieving the required nuclear sophistication is also needed. Nor should the hope for fusion be permitted to detract from energy conservation efforts, from interim emphasis on coal products, or from the long-range benefits of solar resources.[18]

At this writing, there seems no peril whatever that fusion enthusiasm will be in any way to blame if other energy directions are neglected. The general public knows almost nothing about fusion and the educated, reading public knows little. When fusion energy becomes available, as the experts confidently expect, it may come suddenly like a blinding flash of light and inspire a worldwide religious revival, singing in the streets, and a veritable orgy of turning on all the lights and air conditioners and letting them run till dawn.

THE FUSION TIMETABLE

When? We keep coming back to that. At this juncture, the question, of course, can't be answered. The directions to take in reaching the fusion horizon are substantially agreed on. But only educated guesses can be hazarded concerning the estimated time of arrival. However, there is some agreement in these educated guesses by the experts and an aura of confidence that doesn't seem to decrease as time goes on.

Herman Postma, Director of the Thermonuclear Division, Oak Ridge National Laboratory, wrote in 1970:

> Considering the problems and ramifications of the engineering and environmental aspects of fusion, we know we are only at the beginning; nothing has been cast in stone. The main story is still the confinement problem, one that the plasma physicists must solve. Yet, increasingly, the nuclear engineer is getting more involved. Even if the confinement problem were solved tomorrow, the time to solve many of the engineering problems and to produce plentiful and reliable power economically will still be a very long time. Some say 30 years or more. On the other hand, fusion power seems certain to come eventually.[19]

2020 FUSION

By 1973, progress in fusion research brought this statement from the Division of Controlled Thermonuclear Research of the U.S. Atomic Energy Commission:

> Plasma physicists believe that in the coming decade the scientific proof that fusion reactors can indeed be built can be obtained. Given that proof, the fusion program can shift into a development phase in which the practical problems of producing economically competitive electrical power can be fully addressed .... An analysis of what might be accomplished in an orderly aggressive program indicates that central station fusion power might become commercial about the year 2000. Assuming any of various models for its introduction into the utility market thereafter, fusion power could then have a significant impact on electrical power production in the year 2020.[1]

This cautious prediction/anticipation encourages setting our sights on 2020 fusion, less than 50 years away. Considering the familiar identification of normal human vision as "2020," naming that particular year as a conservative goal for large-scale fusion power may seem coincidentally prophetic.

Progress continues with new advances building on those that have gone before. Professor Kammash in 1974 emphasized his conviction "that scientific feasibility of controlled fusion is not far off." "This optimism," he writes, "is based on the much better understanding we now have of plasma physics than we did a decade or so ago," adding that it is "an optimism based on knowledge."[2]

Establishment of scientific feasibility will probably inaugurate a world drive to confront the technological problems that must be solved before power from controlled fusion can be expected.

A convincing demonstration of feasibility could well set off a "Fusion Rush" that would make the Alaskan oil pipeline project, the California gold rush, and the building of the pyramids seem like games for children by comparison. The rush won't start until clear proof is offered that fusion is a dream in the process of becoming a reality. Then watch out. Members of the human race will gather around fusion laboratories and hold their breaths until it happens, because the human appetite for energy is miles ahead of other appetites today, and steadily increasing its lead. According to William Gough, writing for the U.S. Atomic Energy Commission, "If we don't get an unlimited energy source that is relatively inexpensive or at least somewhere close, we will be in trouble. We will be unable to support the large world population at a standard of living anywhere near what we have now in this country or even what less fortunate countries now hope to obtain."[20]

Fusion is a principal aim of that hope, not merely for comfortable living standards but more and more for civilized survival. Mankind will wait for 2020 fusion with impatience, and hope it can come sooner. If fusion fulfills its promise, it will be an unlimited energy with the potential for liberating us from a chilly and shadowy fate.

Note: All fusion references follow Chapter 3, Fusion and the Environment. See p. 38.

FUSION AND THE ENVIRONMENT

"IF I GROW UP"

A much-repeated though not especially funny joke of the atomic age is the story of the little kid explaining "what I'm going to be if I grow up." A switch of prepositions from "when" to "if" speaks volumes about our era; and with the U.S. in 1977 debating a new weapon, the neutron bomb, the likelihood of a permanent return to "when" seems remote.

In 1975 a Nobel Prize winning biologist, George Wald, echoed contemporary fears in a magazine article:

> I still have young children at home; and I teach 200 magnificent young people at Harvard. Though I desperately want to, I cannot find any assurance that they have a future—any assurance that they will be in physical existence ten, twenty, twenty-five years from now. One does not live with that kind of thought day in and day out. One puts it away; one shuts it off and tries to live some kind of normal life. But it is always in the background. We live with a constant sense of the insecurity of life on Earth today.[21]

Professor Wald was disturbed by mounting pollution from fossil fuels, human waste and neglect, stockpiles of indestructible but highly destructive plutonium, and the fear that pervades modern life. Science hadn't given him reassurance. Perhaps the professor should consider the potential human benefits of fusion energy, which might supply greater reason to hope for the better, if he could believe in those benefits.

When the bare facts of fusion are considered, it does sound almost too good to be true, like a storyteller's entertaining fairy tale for children. Fusion energy when it comes may be as "clean" as Snow White.

"FUSION NOT FISSION"

If there is a rallying cry these days for scientific environmentalists, that may be it.

The worldwide petroleum shortage and energy scare during the winter and spring of 1974 added further confirmation to what scientists, sociologists and politicians must have suspected in advance: That men would accept, even choose, environmental poisoning rather than live permanently with an uncomfortable diminution of their normal energy supplies. Many environmental standards were quickly compromised if doing so could increase available energy.

"What matter pollution if my conveyance and my company are adequately fueled to keep me mobile and employed."

With that attitude so prevalent, it is natural and predictable that those most concerned about restoring and preserving the human habitation are among the first in line to herald the advent of fusion. Fusion offers the hope of having our cake and eating it too: Plentiful energy without the pollution perils associated with fission and traditional fuels.

The environmental dilemmas presented by traditional fuels (the fossil offspring) need not be recounted here. They are not long-range energies of the future anyway, and the problems are well known although they have never been faced very aggressively.

Fission reactors may be with us until fusion reactors can replace them, but they are likely to continue inspiring anxiety, apprehension and suspicion. Whatever the safeguards, fission nuclear reactors may always pose the theoretical danger of radiation leakage, atomic accident, and the seemingly perpetual dilemma of radioactive waste products.

The following is from a 1971 report by the Environmental Protection Administration of the City of New York:

Management of High-Level Radioactive Wastes Poses a Major Hazard. Presently, about 80,000,000 gallons of high-level radioactive wastes are stored in liquid form in about 200 concrete-encased steel tanks buried at AEC sites in Washington, South Carolina and Idaho. At the Western New York Nuclear Service Center (WNSC), 30 miles south of Buffalo, in West Valley, New York, about 520,000 gallons of high-level radioactive wastes are stored in liquid form in two 750,000-gallon carbon steel tanks and two 15,000-gallon stainless steel tanks owned by NYASD.

Some tanks are cooled, others are allowed to boil with steam siphoned off to prevent rupture. Aside from the possibility of an accident, there is a clear risk that the tanks will corrode and leak. At best, the tanks are expected to last about 20 years (though the wastes within them remain deadly for 600-1000 years) before requiring replacement to protect against deterioration of protective materials.[22]

Subsequent to that year, different methods for disposing of radioactive wastes have been tried (such as converting liquid wastes to solids and storing in dry geologic formations such as salt mines), but the criticism has not stopped, nor probably have the theoretical dangers. If theoretically something can go wrong with catastrophic results (and a number of such possibilities are associated with fission), the prudent assumption has to be that someday it might.

Hence, one of the key advantages promised by fusion. According to Professor Kammash at the University of Michigan, radioactive waste products are absent from fusion reactors, the danger of nuclear explosions is eliminated, and there is no significant waste heat problem. Safety comes from the essential balance between plasma density, plasma temperature, and confinement time. Disturbing this balance inevitably forces automatic shutdown, eliminating even the possibility of an explosion.

FUSION BENEFITS

The environmental advantages of fusion also include: "Low biological hazard in the event of sabotage or national disaster: Again, if you assume the p-Li^6 cycle we have no hazard. For the

D-He^3 cycle we have practically no hazard because there are no volatile radioactive materials or fuel. For the D-T fuel cycle you must compare the biological hazard for the possible release of tritium to the atmosphere with say volatile fission products from nuclear systems, and there seems to be a very wide margin in favor of fusion."[20]

With some fuel cycles such as D-T (Deuterium-Tritium), which may well be supplanted as fusion technologies advance, there is a radioactive fuel: Tritium.

In fusion reactors using tritium, wall radioactivity will be induced requiring periodic changing of the container walls. But according to Gough, the radioactivity imposed by tritium is "still many orders of magnitude less hazard potential than is present in a fission system."[20]

Transport of fuels offers a further safety edge for fusion over fission, even when tritium is involved, since only nonradioactive lithium and deuterium require shipment regularly, while the plant will breed its tritium on site.

Ultimately, fusion reactors are expected to rely on fuel cycles, such as D-D (Deuterium-Deuterium), with no radioactive waste products, inherent safety against nuclear explosions, a minimum after-heat problem, and relatively low waste heat.

Waste or unutilized heat distributed into the immediate environment is another drawback of fission reactors avoided by fusion reactors. With as high as 90% efficiency, direct conversion fusion reactors reduce waste heat to the point where they can be comfortably located in or close to heavily populated areas.

> Lower efficiency plants could also be located in the center of a city if a use for the waste heat could be found . . . such as the heating of buildings or the distillation of sewage.[20]

All available data on fusion tend to emphasize the environmental safety aspects of the process. Experts seriously emphasize the logic of establishing fusion reactors in metropolitan areas. There is no danger to the population, and the costs of transmitting energy to users is reduced. A runaway reaction is impossible because of the nature of plasmas and because there is approximately only one gram of fuel in the container core.[1]

Will all of these promises be fulfilled or not? Only time knows the answer and will in due course make the answer plain.

> Nuclear power, perhaps more so than any form of energy, can be a great blessing or an awful curse to mankind in the future. The basic objective of the nuclear energy program is to provide energy that is cheap enough and plentiful enough so that it becomes a basic raw material. The present light-water reactors are but the first commercial step along that road. If the dreams and aspirations of the nuclear scientists and engineers can be achieved, the abundant supply of truly low-cost energy from the breeders and fusion power will provide answers to the pressing shortages of food, water and metals here and throughout the world. They could move mankind into a new era of material abundance.[23]

In terms of both energy and environmental needs, fusion alone appears to be the answer. Other types of nuclear energy providers may have the capacities to deliver energy, yet many people consider the assault they threaten on the environment too great to accept without anxious reservations. If the price of fission is a constant concern, the price may be excessively high. Men do not live easily with growing tigers. The ideal, of course, is an energy source that does not growl, that grants mankind the luxuries of mental serenity and physical safety. Fusion could be that energy source.

> The end products of the fusion reaction—helium, hydrogen and neutrons—can hardly be better chosen from the point of view of easing environmental pollution. It has even been proposed that the neutrons be used to clean the environment by transmuting certain particularly dangerous long-lived radioactive by-products of fission reactors to harmless stable nuclei.[11]

It is an aging dream—a plentiful energy that does no damage, only good. Hopefully fusion will make the dream a reality, and prove itself virtually as clean as sunlight. Making such dreams come true is the exclusive challenge and magnificent opportunity of science. Fusion, of course, is a relatively humble dream. Scientists are asked only to explore and to master the interiors of stars.

FUSION REFERENCES

1. "Fusion Power: An Assessment of Ultimate Potential," Division of Controlled Thermonuclear Research, U.S. Atomic Energy Commission, February 1973.
2. Kammash, Terry. *Fusion Reactor Physics: Principles and Technology* (Ann Arbor, Michigan: Ann Arbor Science Publishers, 1975).
3. Rose, D. J. "Controlled Nuclear Fusion: Status and Outlook," *Science,* Volume 172, Number 3985, May 21, 1971, pp. 797-808.
4. Post, R. F. and F. L. Ribe. "Fusion Reactors as Future Energy Sources," *Science,* Vol. 186, No. 4162, November 1, 1974, p. 397.
5. The National Energy Plan, Executive Office of the President, Energy Policy and Planning, April 29, 1977, p. 78.
6. Gough, W. C. and B. J. Eastlund. "The Prospects of Fusion Power," *Scientific American,* February 1971, pp. 50-64.
7. Coppi, B. and J. Rem. "The Tokamak Approach in Fusion Research," *Scientific American,* July 1972, pp. 65-75.
8. Metz, William D. "Nuclear Fusion: The Next Big Step Will Be a Tokamak," *Science,* Volume 187, No. 4175, February 7, 1975, p. 423.
9. Glasstone, S. "Controlled Nuclear Fusion," Understanding the Atom Series, Division of Technical Information, U.S. Atomic Energy Commission.
10. Gough, W. C. "Fusion Energy and the Future," *The Chemistry of Fusion Technology,* D. M. Gruen, Ed. (New York: Plenum Publishing Corporation, 1972).
11. Lidsky, L. M. "The Quest for Fusion Power," *Technology Review,* January 1972, pp. 10-21.
12. *Fusion Forefront,* ERDA-CTR Newsletter, Vol. 8, No. 3, December 1975, p. 1.
13. *Ibid.,* p. 2.
14. Metz, William D. "Economic Breakdown May Be the Problem for Laser Fusion," *Science,* Vol. 186, No. 4170, December 27, 1974, p. 1194.
15. *Ibid.,* p. 2.
16. Metz, William D. "Energy: Washington Gets a New Proposal for Using H-Bombs," *Science,* Vol. 188, No. 4184, April 11, 1975, p. 136
17. Ashworth, Clinton P. "A User's Perspective on Fusion," Presentation at Atomic Industrial Forum, "Conference on Energy Development: Probing the Possible," February 11, 1976.
18. Parsegian, V. L. "The Future of Fusion," *Science,* Vol. 187, No. 4173, January 24, 1975, p. 213.
19. Postma, H. "Engineering and Environmental Aspects of Fusion Power Reactors," *Nuclear News,* April 1971, pp. 57-62.
20. Gough, W. C. "Why Fusion? Controlled Thermonuclear Research Program," Division of Research, U.S. Atomic Energy Commission, WASH 1165, June 1970.

21. Wald, George. "There Isn't Much Time," *The Progressive,* Vol. 39, No. 12, December 1975, p. 24.
22. Fabricant, N. and R. M. Hallman. *Toward A Rational Power Policy— Energy, Politics, and Pollution* (New York: George Braziller, 1971) pp. 139-141.
23. Freeman, S. D. "Policies Affecting the Energy Needs of Society," *Nuclear Power and the Public,* H. Foreman, Ed. (Minneapolis, Minnesota: University of Minnesota Press, 1970), pp. 168-178.

CHAPTER 4

SOLAR ENERGY

"We move from the complex to the simple, and
the obvious is the last thing we know."

Elbert Hubbard

REMARKABLE GIFTS FROM A MIDDLE-AGED STAR

Worried about running out of oil, coal, shale, natural gas, uranium, lithium? Concerned about turning up the thermostat to warm your bones, and nothing happens? Take a look around; or more precisely, if it is clear tomorrow, face east during the morning and glance at the sky. That brilliant object overwhelmingly dominating the view is called the sun.

During the past 4.5 billion years, while the earth has been developing centipedes, human beings, energy shortages, and the Grand Canyon, the sun has continuously donated an endless stream of gentle, nonpolluting, inexhaustible energy. The very existence of life on earth is proof that the sun's energy has been remarkably potent and effective, despite the fact that only about 65% of the energy received is absorbed and only a minute portion is directly utilized by man.

There is no doubt, however, that solar energy earns full credit for life, weather, climate, and the ecological character of the earth. The amount of that energy is vast, an estimated 25,000 times greater than man's entire current energy production on earth via other sources.[1] Indeed much of man's own energy production came originally from the sun millions of years ago (fossil fuels). The sun has always been man's foremost energy source, and it always will be. If the sun stopped, man too would stop.

> In terms of *total* energy the main energy source for any society is the sun, which through the cycle of photosynthesis produces the food that is the basic fuel for sustaining the population of that society.[2]

Astronomers and astrophysicists are clever people with sensitive apparatus, and they tell us that our sun is a middle-aged star, 860,000 miles in diameter, functioning as a thermonuclear fountain of colossal energy. Every second 630 million tons of hydrogen are converted to helium. During that same second, nearly 5 million tons of hydrogen are converted into solar energy. This is the energy that journeys across the 91,377,000 miles (147,053,000 kilometers) separating earth and the sun at the point of perihelion. This is the energy that makes human life and all other life possible on earth.

Not only is life made possible by energy from the sun, the earth itself and the chemical elements that compose it had their birthplace in the sun. By a process called nucleosynthesis, all the elements are produced from hydrogen by means of thermonuclear reactions in the awesome interiors of stars. In our case and our earth's case, this magic transformation was accomplished by the sun.

Scientists agree about this, but at the moment they are not 100% certain about the "how." Scientists never sit still permanently, and one of their main jobs is questioning everything. So as you might expect, questions are now rising concerning the traditional view of what happens inside a star such as the sun to produce thermonuclear deluges of energy.

TEN TO TWENTY MILLION DEGREES CELSIUS

Science continues to probe, and new facts are learned while older "facts" are challenged. Some current facts about the sun are that its temperature is between 10 and 20 million degrees Celsius, that its density is 100 times that of water, that its pressure exceeds 1 billion atmospheres. Those are just big numbers, of course. We nod and want to know what else is new.

What's new is a contemporary doubt that the explanation of solar and stellar energy generally accepted during the past decade is the whole story. This explanation involves two cycles of nuclear reactions sustained by the high temperatures present in the sun's core. Hans Bethe won the Nobel Prize for Physics in 1967 for his work in this field.

Stellar energy was thought to result from a chain reaction known as the carbon cycle (also called the carbon-nitrogen-oxygen, or CNO cycle) and the proton-proton cycle. The function of each cycle, according to theory, is bringing about the conversion of hydrogen to helium with a release of high-energy radiation. This conversion was briefly discussed in the chapter on fusion. In the carbon cycle, carbon serves as a catalyst for the conversion of hydrogen in a complex chain reaction. In the proton-proton cycle, by many considered to be the chief source of stellar energy, two hydrogen nuclei (protons) ($_1H^1$) undergo thermonuclear fusing to form a deuterium nucleus ($_1H^2$). Another hydrogen nucleus or proton is fused with the deuterium nucleus to form an isotope of helium ($_2He^3$), which in turn fuses with another helium isotope, forming an ordinary helium nucleus (two protons and two neutrons) and releasing two protons plus vast amounts of fusion energy.

Elementary, my dear Watson. Sounds quite simple doesn't it, and this has been the generally accepted explanation of what goes on inside stars to give careless people sunburns at the beach on Saturday afternoon. The explanation makes mathematical, physical and chemical sense. It has a logical ring. But is it true?

Some scientists wonder. They point out that if the sun's energy results from the fusion of hydrogen nuclei in the core, with the formation of helium and the release of energy, there should be simultaneous production of neutrinos (subatomic particles without mass and electrical charge). But solar scientists have not been able to detect neutrinos. Until they do detect them, the traditional theory is suspect.

A counter theory has been suggested by Dr. J. Christensen-Dalsgaard and Dr. Douglas O. Gough at Cambridge University. They have evidence that the sun may pulsate, expanding and contracting approximately 5 miles every 2 hours and 40 minutes. If confirmed, this fact would require changes in the solar model that might not accommodate the theory explaining solar energy as a product of fusion. An alternate hypothesis: Solar energy comes from a "black hole" or superconcentrated energy source at the sun's core.

Whatever the explanation proves to be, the sun regularly comes through with its vital energy deliveries, in sufficient quantities to keep the earth warm enough for life and with vast amounts of energy left over if we could find efficient ways to capture and use it.

CONSTANT OR VARIABLE?

Is the sun's energy constant? It was thought to be, but this too is now questioned.

> What is now being suggested is that the sun, far from being the constant star of recent memory and astronomical theory, has in the past 1000 years undergone several significant changes in its magnetic activity and, perhaps, in its output of energy. If so, then future changes in solar activity cannot be ruled out.[3]

Records suggest that the sun at the end of the seventeenth century behaved differently than it behaves now. Factors indicating changes include sunspot and auroral activity, shape of the solar corona, carbon-14 (^{14}C) concentrations in the earth's atmosphere, and the earth's surface temperature.

Even tree rings teach us more about the sun. Consider carbon-14. This is formed in the earth's atmosphere by cosmic rays and appears everywhere on earth. Tree rings offer a radiochemical diary of ^{14}C. If solar magnetic activity is intense, cosmic rays are partially blocked and ^{14}C in tree rings for the period is reduced.

Practically everything occurring on earth tells us more about the sun. "Someday," Thomas Edison predicted, "we will harness the rise and fall of the tides and imprison the rays of the sun." Every living creature, of course, continuously imprisons the rays of the sun. We feast on sunlight one way or another every second of our lives.

The sacred fires in the Greek temple at Delphi were lighted with solar energy focused through mirrors. The Greek scientist Archimedes allegedly drove a fleet of Roman ships away from Syracuse in Sicily by means of a burning mirror reflecting sunlight. There is the legend of Icarus whose father Daedalus made the boy wings from wax and feathers; but Icarus flew too near the sun and fell into the sea when the wax melted. The same as any other energy, solar energy isn't always a feast. The same as any other energy, it needs wise using. There is so much of it, if we could effectively use even a minute portion, our energy worries would be much reduced.

MORE JOBS FOR SUNLIGHT

Is there anyone or anything to prevent us from taking greater advantage of solar energy in the future? Only ourselves. It is here daily, ready to work, doing all its regular jobs with the effortless familiarity of several billion years, and quite willing to add on further tasks.

The figures can be startling if you have never considered them before:

> In 1970 the total energy consumed in the U.S. was about 65 x 10^{15} Btu, which is equal to the energy of sunlight received by 4300 square miles of land, or only 0.15% of the land area of the continental U.S. Even if this energy were utilized with an efficiency of only 10%, the total energy needs of the U.S. could be supplied by solar collectors covering only 1.5% of the land area, and this energy would be supplied without any environmental pollution. With the same 10% utilization efficiency, about 4% of the land area could supply all the energy needs in the year 2000. By comparison, at present 15% of the U.S. land area is used for growing farm crops. For some applications, such as heating water and space heating for buildings, the utilization efficiency can be much greater than 10%.[4]

These figures, of course, have to be taken with a grain of economic salt. There are reasons why all these possibilities haven't been turned into realities, and probably the chief reason is economics. Other energies have been cheaper to use.[5] Solar energy undoubtedly will be used increasingly where it is economical to do so, and for many applications that has already happened.

Solar energy until the recent past had not been extensively exploited because technologically it had not proven competitive in terms of price and convenience with traditional fuels. The situation is changing, however, as traditional fuels rise in price and become rare. A parallel development is the continuing improvement in the technology of solar energy.

According to J. Richard Williams, Associate Dean for Research at the Georgia Institute of Technology, in his book *Solar Energy,* this energy is ready now to supply about 25% of U.S. energy demands economically.[4]

In June 1974, a spokesman for the U.S. General Accounting Office said, "American ingenuity and the soaring prices of traditional fuels may make the sun's energy a bargain for the average home before too long." The G.A.O., more noted for fiscal despondency than for romanticizing its figures, informed the U.S. House of Representatives Committee on Science and Astronautics that its analyses showed solar heating and cooling can be made competitive.

The key point about solar energy could be that it is available, ready for the taking, permanent. This fact should help spare us extreme anxieties and psychoses as energy depletion spreads in traditional fuels. The Queen of France's advice for the people to eat cake when they had no bread can be relevantly paraphrased for our energy future: If we have no traditional fuels to consume, let us consume sunlight.

The use of solar energy for a wider range of purposes than mental contentment in the spring and that tropical look after two weeks on the beaches of Guadeloupe is beginning to happen.

The most practical and obvious immediate application of solar energy is in the heating and cooling of buildings, heating water, and supplying concentrated heat for drying functions in industry and agriculture. Proponents note that in addition to these services, it will also be practical soon to employ solar energy on a large scale for the pollutionless manufacture of electric power.[4]

THE SOLAR REASSURANCE

Have you noticed how both experts and laymen respond to the energy challenges confronting the human race in the final quarter of the twentieth century? Referring to fossil fuel shortages, there is a note of urgency in their comments; but there rarely seems to be panic. Practically everyone is convinced that technology will supply reasonably abundant alternatives to the fossil fuels. The mysteries of fusion will be untangled. Or the problems of nuclear fission will be solved. *Something* will turn up to keep the lights burning and the cars running. But faith in technology might not be sufficient to avoid greater alarm, if not panic, except for one conspicuous energy reality that reassures us daily: The sun is still in business at the same old stand.

Ask any expert about energy—each of us is an expert these days—and you'll receive a ready dissertation on the expert's pet energy. Fusion is the answer, he'll say and explain why. In the long run, geothermal energy from mother earth will warm our bodies and cook our stew, he'll argue, with handy facts and figures. Listen, he'll whisper conspiratorially, we'll get all the energy we need from fermenting strawberries, or extracting chlorophyll from leaves, or domesticating herds of fireflies to provide illumination. Something. But whatever the expert's pet theory about man's energy future, you can always sense that periodically he glances gratefully toward the sky. If the fireflies don't come through, there's always the sun.

"Case money" is a popular American slang expression for a person's emergency funds, his nest egg, his special reserve for a time of great need. Obviously, we tend to think of the sun as a permanent and dependable source of energy, our "case energy" that we can fall back on if nothing else works out.

47

To this point, the solar reassurance has been perhaps the main application made of the solar possibility. Its utilization as a substitute for disappearing fossil fuels has been limited so far. In 1976, it was reported that solar energy supplied less than 0.5% of U.S. energy requirements, and this figure classified the burning of wood for heat as a solar energy category.[6] This same report concluded that solar energy because of technological and economic factors probably would not contribute more than 1% of total U.S. energy needs in 1985; but it saw bigger days ahead for solar energy:

> Progress in dealing with the technical, economic, and environmental factors . . . will, in our opinion, lead to a "coming of age" of solar energy in the years between 1985 and 2000. We expect that in the period after the year 2000 solar energy will have become one of the conventional energy sources used in many regions of the world. However, attainment of this eventual success will demand patience and a continued dedication to the advancement of solar energy technologies in the intervening years.[6]

Question: What is now being done to advance solar energy technologies? Answer: So much we have to wonder if the years 1985-2000 aren't more likely to be fulfillment years for solar energy, with a precocious "coming of age" happening sooner. In some applications, solar energy already seems on the verge of widespread adoption by smart people who can see the handwriting on the wall.

SOLAR ENERGY PROJECTS

Three broad applications of solar energy have been identified as those with optimum promise in terms of technical and economic realities. They are:

1. heating and cooling residential and commerical buildings;
2. chemical and biological conversion of organic materials to liquid, solid and gaseous fuels;
3. generation of electricity.

Everyone from enthusiasts to sceptics agrees that solar energy shows legitimate promise of making a significant contribution in each of these areas. Whether or not solar energy is allowed to fulfill its promise as quickly as it might involves a tangled complex of factors. Will government and private funds be sufficiently invested to perfect technologies? Will shortages of other energies give impetus to the solar initiative, or will temporary surpluses of exhaustible energies inhibit solar progress with a period of idling or do-nothing delays?

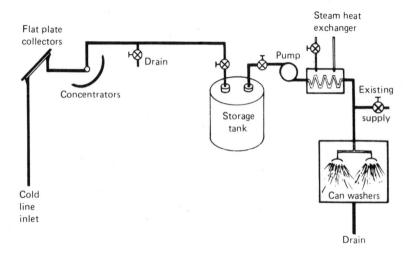

Figure 2. Solar energy for process heat in a canning operation. Acurex Aerotherm (Mountain View, California) in association with the National Canners Association is testing the use of solar energy to produce hot water needed in food processing. The diagram shows the essentials of the system, utilizing a closed-loop tracking unit to follow the sun and aluminum surface collectors with glass-enclosed receiver tubes. This is one of many projects aimed at broadening the application of solar power to industrial needs.

In the 1977 U.S. National Energy Plan, it was reported that "Solar hot water and space heating technology is now being used, and is ready for more widespread commercialization. A temporary federal program of financial incentives and public education is needed to stimulate the development of a larger solar market." The plan offers this related proviso: "The Government should aggressively promote the development of nonconventional resources despite the fact that they face many uncertainties. The danger of too much initial skepticism is that it may become a self-fulfilling prophecy."[7]

But if popularity and public interest are a relevant guide, solar energy should have less and less trouble finding more and more practitioners. Discussing the colonial situation with a British Royal Governor, Benjamin Franklin once said that British authorities should be less concerned with what people *should* do and more concerned with what they are *apt* to do. This is a truism in connection with solar energy, and what people seem apt to do in growing numbers is to try it.

In the second half of the 1970s "trying it" was clearly becoming the thing to do. ERDA was coming through with extensive research funds. Solar energy was being applied in more places and more ways than ever before. J. Richard Williams wrote that technological breakthroughs have produced new types of solar collectors and more efficient coatings for operation at the higher temperatures required for air conditioning.[4] This is written shortly after the New York City power failure and blackout of July 13, 1977. The blackout was blamed in part on the tremendous energy service demanded of the power system under summertime conditions.[8] Clearly, the wider use of solar collectors and storage facilities during summer months would lessen peak demands on traditional power sources and reduce the danger of blackouts. In 1977, if New York City and other cities were not yet prepared for a major commitment to solar power, some citizens in those cities were paying attention. At least one residential building in New York had prepared in advance for mammoth citywide power failures by taking advantage of solar energies, both wind and sunlight, with the aid of federal funds.

> Federal funds paid for a $4000 windmill erected on a residential building in New York City. The windmill with a 2000-watt windpower generator attached was expected to produce electricity to light the hallways and supply 85% of the building's

hot water. The same building had been equipped with water-heating solar collectors to make it as energy self-sufficient as possible.[9]

Other city buildings, however, with previously announced intentions of moving ahead in solar developments had been frustrated by difficulties in obtaining timely deliveries of solar collectors. In December 1975, the RCA Corporation at Rockefeller Center shelved plans to install solar apparatus on its Rockefeller Plaza building, and cited the problem of delivery. For the same reason, the First National City Corporation in New York was reported to have dropped plans for equipping a new skyscraper with solar heating and cooling.

SOLAR ENTERPRISE

Yet the very fact these plans had been made were an indication of the progress in solar technology, and they suggested that eventually buildings in cities such as New York would be equipped to take advantage of the solar promise.

> Grants from the federal government have supported the construction of a variety of solar homes and buildings, as well as the retrofit of existing facilities for solar heating and cooling . . . . The total number of solar homes built and under construction in the U.S. has escalated rapidly over the past two years, and as of 1976 there are at least 1000 solar homes under construction or completed, both with federal and private support. The demand for solar collectors and other solar systems components has stimulated manufacturing companies to make solar hardware available. Only a few years ago commercially manufactured flat-plate solar collectors, differential thermostats, pumps for home solar water heating applications, and solar air conditioning units were difficult to obtain within the U.S. Now these products are readily available from a variety of manufacturers.[4]

The number of homes and buildings actually being fitted with solar energy facilities was not great in 1977, but it was a steadily growing number; and every successful application served as a prototype for hundreds and later thousands of applications in the future. Mobil Corporation in a 1977 newspaper advertisement said that "Solar energy on a significant scale is coming, no doubt about it. But how many decades hence is still a question."

The question was being answered with both predictions and plans by those in a position to know. An ERDA official in June 1977 reported the energy agency's estimate that 10% of U.S. energy needs could be met by solar energy in the year 2000.[10] In 1975 ERDA's estimate was only 7%. A total of 1.3 million solarized houses was named as a national goal by 1985. Other plans to solarize federal buildings were also going forward. The General Services Administration indicated that it would be "standard practice to consider the feasibility of including a solar energy system as a part of the early design process of each new building project. In cases where a solar system is found feasible, it will be included as a part of the original construction if funds are available."

All the solar interest and action in addition to promoting development and intelligent use of earth's oldest energy also meant something else, and on a large scale: Business.

The scope for future enterprise was forcefully suggested by a 1976 article in *National Geographic:*

> By the year 2000 today's dawning solar technologies may have become a 25-billion-dollar-a-year industry . . . . Most estimates agree that in 25 years solar systems could save more barrels of oil than will be flowing through the Alaskan pipeline—or about a third of all our current imports. That amounts to several billion dollars a year in balance-of-payments savings. And as one lawmaker recently noted, "Sunshine cannot be embargoed."[11]

There doesn't seem any doubt about it. After years of playing second fiddle, or thirty-second fiddle, in the energy picture, the sun at last seems on the verge of becoming a star.

A newspaper typographical error helps us keep the sun in accurate perspective: The *New York Times* reported on sun flare studies by the Space Environment Laboratory at Boulder, Colorado, indicating that the area of an "energetic sun flare" was around 4 million square miles! Sun flares, significant in ways not yet fully understood to terrestrial climatic events, are

52

associated with 11-year sunspot cycles. The flares give off X-ray radiation and in the ionosphere serve as a rebounding board for radio waves. The *Times* later printed a correction: The area of solar flare activity, reported as 4 million square miles, was actually 400 million square miles.

It is clear that this energetic star is going to be given many different roles to play in an effort to satisfy the energy needs of the years ahead.

THE NEW DO-IT-YOURSELF BONANZA: SOLARIZATION

Many solar systems for heating water and even home heating are designed and marketed for people with mechanical skills and a do-it-yourself bent to perform much of the installation work themselves. J. Richard Williams supplies details on "A Solar Heating System You Can Build Yourself."

According to Williams, "a solar heating and domestic hot water system can be installed by any individual sufficiently skilled in carpentry, plumbing, and electrical work to add a room, finish a basement, or perform similar construction and repairs. The required electrical work involves some low-voltage wiring. The pumps and control units can be plugged into existing electrical outlets if desired. Plumbing techniques are no more sophisticated than those required to install a sink. The solar collectors can be built into the roof or mounted on the roof, or located on the ground, over a patio, or any other convenient location that is unshaded most of the day."[4]

Many companies, more all the time, are ready with equipment to serve the do-it-yourself urge. Most companies selling solar collectors and other accessories of solarization, typically use the sales argument that the installation will pay for itself in energy savings in a few years.[12] These claims may be accurate, but many factors have to be taken into account when choosing an economical solar system for a particular task. "Sizing the array" or calculating the collector arrangement necessary for the desired result must be done carefully. Two basic rules that would seem to be pertinent when considering a solar system for yourself are: (1) Learn what can be done. (2) Know what you're doing.

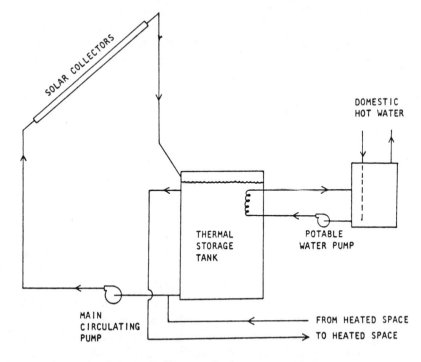

Figure 3. Schematic diagram of solar energy system.

SOLAR ENERGY COLLECTORS*

As the name implies, Solar Energy Collectors collect, concentrate and deliver solar energy. A number of different types have been developed, and new advances are being made. The ideal shapes and materials for effective performance in a broad variety of applications are being studied in both university and government research laboratories. For best performance, the type of collector used is necessarily dictated by the job to be done.

*Much of the information on solar energy apparatus and systems in the following pages was adapted from J. Richard Williams' *Solar Energy: Technology and Applications, Revised Edition* (Ann Arbor Science Publishers, Inc., 1977). Materials from other sources are specifically referenced.

Flat-Plate Collectors
(Low-Temperature, No Concentration)

These are used to heat water and buildings and provide temperatures approximately 150°F (65°C) above ambient. Figure 4 shows the components of a flat-plate collector. A black plate covered by a transparent cover plate or plates is supported by an insulated box. The black plate collects heat from sunlight and transmits the heat to a flowing liquid, commonly water. The heated liquid is used then to circulate heat in a house or building.

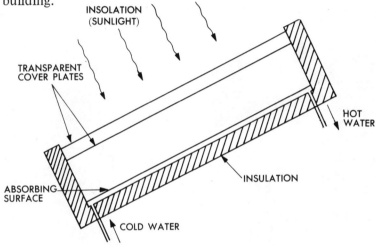

Figure 4. Flat-plate solar collector.

Increasingly sophisticated and elaborate flat-plate collectors are being introduced. NASA scientists have designed a flat-plate collector with a new selective coating absorbing over 90% of solar radiation, and losing only 6% through reemissions.[13]

Parabolic Cylinder Concentrating Collectors
(Medium-Temperature Concentration)

Parabolic Concentrator with Multiple Small Mirrors
(High-Temperature Concentration)

These solar concentrators are used when higher temperatures are needed for electrical production, industrial drying, etc. Solar concentrators generally employ parabolic reflectors to focus sunlight on the collector. The reflector normally must be turned toward the sun. This complication is avoided in simpler, less costly to operate flat-plate collectors. The parabolic reflector concentrates sunlight on the collector. A number of simple to intricate variations of reflector-collector arrangements are available, and for the highest temperatures an intricate arrangement of small flat mirrors focused to the same point is used.

Figure 5 shows a solar furnace at Odeillo Font Romeu in the French Pyrenees. The sun shines here 180+ days per year, and

Figure 5. French solar furnace (Courtesy of J. D. Walton).

1000-watt solar intensities per square meter are reportedly achieved. The parabolic reflector contains 9500 mirrors in its 130 ft x 175 ft (40 m x 53 m) surface. It is fixed while eight tiers of heliostats follow the sun. The French solar furnace was completed in 1970 at an estimated cost of $2,000,000.

A smaller furnace of the same basic type in the Soviet Union is said to melt refractory materials at temperatures reaching 3500°C.[14]

SOLAR ENERGY APPLICATIONS

The full potential contribution of solar energy has barely been tapped, yet worldwide efforts by researchers, designers and builders, combined with widespread public interest are proofs of vitality. Many specific current applications can be mentioned. The list grows significantly when future applications are included.

Solar Water Heaters

Already used in many areas of the world, these provide a simple means, at locations with adequate sunshine, of heating water. Solar water heaters are sold commercially in Australia, India, Israel, Japan and the USSR. Several million solar water heaters are regularly used in Japan, Israel and other countries. The application could be beneficial in many more areas, including vast portions of the U.S.

In a typical arrangement, a flat-plate solar collector on the roof provides hot water by natural circulation to a tank higher on the roof. The mechanism is a practical way to save conventional fuels. Simple water heaters of this type are inexpensive, but for optimum effectiveness require more insolation than some latitudes enjoy during winter months. Closed pipe collectors and other elaborations of the basic system have been developed to increase effectiveness during periods of reduced sunlight. In most parts of the U.S., buildings of all sizes can rely on solar energy for significant portions of their water heating requirements. Williams estimates that approximately 60-70 ft^2 of collector area could supply three-fourths of the water heating needs in U.S. apartments.

Colder areas benefit from special closed-loop and natural circulation systems, or combinations of solar heating and traditional energy sources.

Many companies sell solar hot water heaters in the U.S., with prices ranging from a few hundred dollars to as much as $2000. Use of solar energy to produce hot water is considered the easiest and therefore the first step toward full utilization.

Figure 6. This solar home was designed by Innovative Building Systems under a U.S. Department of Housing and Urban Development grant. A solar-assisted heat pump provides space and water heating in a 2000-ft^2 residence. A liquid circulating solar collector (700 ft^2) functions through a heat pump to energize a 2000-gallon storage tank. The system is expected to supply 70% of space heating and 55% of water heating requirements annually.

To achieve greater efficiency in the production of hot water as well as other solar energy applications, research efforts are being concentrated on the development of better collectors. Various new designs have been tried, and more are on the way. Such innovations as honeycombed areas, evacuated spaces, and the use of different construction materials combined with concentration and tracking mechanisms are aimed at reducing thermal and optical losses. Such work has made significant progress, and there is no reason to believe it won't continue to effect real improvements.[1 5]

Solar Air Conditioning

A thermal absorption–type refrigeration system operated by solar heat is becoming a fairly common means of cooling. The system operates similarly to ordinary electric refrigerators, except that ammonia is given a high vapor pressure via solar heating rather than mechanical compression.

Air conditioning systems based on absorption through the use of solar energy have been operating successfully for years.

As in many other fields of solar energy application, the challenge involves both technology and economy. Solar cooling technology is available, but not yet at a cost competitive with conventional air conditioning approaches. Getting the price down is a key aspect in most phases of solar energy expansion. Again plenty of work is being done to make solar cooling both competent and competitive. In addition to absorption refrigeration, other avenues are being explored including such methods as adsorption, Rankine/vapor compression, and nocturnal radiation. Integrating solar cooling with solar heating is the main thrust of current efforts because most users need both.[1 6]

Combined Solar Heating and Cooling Systems

Since most houses and buildings need year-round cooling and heating systems, combined solar systems are now available. The solar collector, necessary for either heating or cooling can

60

actually be more efficiently employed when it is used for both. Combined heating and air conditioning systems have proved economical and practical in the U.S. and many parts of the world. With shortages and economic pressures raising the cost of traditional fuels, the appeal of solar energy systems for heating and cooling dramatically increases. That solar energy technology is ready for the challenge is confirmed by NSF and NASA.

One example of solar success is a house designed and built by Harry Thomason in a suburb of Washington, D.C. Harvard physicist William Shurcliff described the house as "irritatingly simple and strikingly successful." The house utilizes a roof of plastic- and glass-encased corrugated aluminum, insulated underneath and painted black to absorb sunlight. A pipe system of pump-powered circulating water is the heat carrier. Over a number of years Thomason's auxiliary fuel costs have been negligible. In 1974, the house was reported to obtain 80% of its heat from solar collectors, and to have a total fuel bill for three years of $18.90!

Dr. Thomason, holder of many solar patents and president of Thomason Solar Homes in Washington, D.C., became interested in solar energy when he took refuge from a sudden storm in a North Carolina barn. Warm rain dripping through the roof started him thinking about a running water system for a solar-heated house.[1 7]

Dr. Thomason is contemporary proof of Emerson's or Hubbard's (both were credited) suggestion that the man who builds a better mousetrap will find the world beating a path to his door. Thomason's house brought the Subcommittee on Energy from the U.S. House of Representatives to his door. The impressed subcommittee left to sponsor legislation providing $50 million for further research on solar heating and cooling. At 1974 costs, Thomason estimated his system added about $2000 to the cost of a four-bedroom house. Fuel savings pay for the system in no more than ten years.

Dr. Thomason manifestly does not have rocks in his head, but he does have three truckloads of rocks in his basement. The rocks surround the 1600-gallon water tank and are used to absorb heat. No doubt they make an interesting conversation piece for guests: "Here are my rocks. How do you like them?"

Rocks and the sun and Thomason's ingenuity add up to an elementary but highly functional heating and cooling system. America's housing future, perhaps the world's, is alive and summer cool in Dr. Thomason's Washington suburb.

Life in Sun Houses

The Thomason Solar House is far from being the only success story in this field, and the list is growing steadily as others glance toward the energy-lavish sun and say "Maybe?" Experimental houses have been built in various parts of the U.S. since the 1940s, and evidence has been slowly assembled concerning the efficiencies of solar installations under different climatic conditions. The University of Florida, the University of Delaware and the University of Arizona have been among the academic communities erecting solar buildings. There is an office in Princeton, New Jersey, and the Paxton office building in New Mexico.

Near Tucson, Arizona, the Decade 80 Solar House derives all of its heat and hot water from solar collectors on the copper roof. With air conditioning required 10 months of the year in the Sonoran Desert, the house obtains 75% of its air conditioning from solar collectors. This is considered impressive efficiency and highly promising for a part of the U.S. where solar radiation is especially plentiful.

There are the famous Löf houses in Colorado. Dr. O. G. Löf, Director of the Solar Laboratory at Colorado State University, built a solar-heated house for his family in 1958. He used air as the heat transfer medium and a pebble bed for heat storage. After 18 years in the Denver house, Mrs. Löf said, "I am utterly unaware of this being a solar house." With 600 ft^2 of rooftop solar collectors, the five-bedroom structure obtained all its summer hot water needs and one-third of its winter heat and hot water needs from solar energy.[11]

Figure 7. This is the Harry Thomason SOLARIS home (number 6) using a water trickle collector. Hot water from the collector is carried by gravity to a basement tank where it is stored for use in space heating.

63

Figure 8. This solarized modern home utilizes the Thomason SOLARIS system. It is the Peter Wood residence in Colorado Springs, Colorado.

Figure 9. Artist's drawing of a major solar project directed by Douglas Balcomb, Los Alamos Scientific Laboratory, for the National Security and Resources Study Center at Los Alamos, New Mexico.

Figure 10. These vertical collectors face south to receive maximum sunlight during the winter. Heat is stored in sodium sulfate decahydrate eutectic. The building is an office in Mead, Nebraska, designed by James Schoenfelder of Hansen Lind Meyer Architects.

An Albuquerque, New Mexico, solar house uses movable solar collectors in the outside walls. Facing south, the panels can be lowered to gain full impact from the sun's rays, and then raised at night to hold the collected heat. The collectors are in front of a tier of water drums, which serve as part of the wall and which store solar heat collected during the day. Water for the drums is obtained from a well operated by a windmill (another solar energy device).

In a summary of the heating/cooling prospects for solar energy, the directors of the Solar Energy Laboratory at the University of Wisconsin have noted that thermal energy for buildings represents nearly one-fourth of U.S. energy use. They indicate that such energy can now "readily be delivered from flat-plate solar energy collectors, and the solar energy incident on most buildings is more than adequate to meet these energy needs." They contend that solar costs are moving now toward a competitive level with other energy sources.

> As fuel costs rise and as the supplies of low-cost natural gas become increasingly more difficult to obtain, solar energy will become more competitive and optimum fractions of annual loads to be carried by solar energy will increase . . . . Solar energy for buildings can, in the next decade, make a significant contribution to the national energy economy and to the pocketbook of many individual users . . . . In our view the technology is here or will soon be at hand; thus the basic decisions as to whether the United States uses this resource will be political in nature.[18]

At the present stage of solar technology, however, the cost of solar installations depends to a large extent on where you live, since where you live determines the size an installation must be to deliver required heating and cooling effects. Even as costs are reduced, this disparity based on geography is likely to continue. Williams writes that "collector size for minimum solar heat cost for a 25,000 Btu/degree day house in six locations was found to range from 208 ft^2 (Charleston, S.C.) to 521 ft^2 (Omaha, Nebraska), corresponding to 55% of the respective annual heating loads. In Santa Maria, California, a 261-ft^2 collector can supply 75% of the annual heat requirement."[4] Yet even in the more temperate and colder parts of the U.S.,

Figure 11. An ERDA project begun by the Southern California Gas Company is this specially designed house described as the Minimum Energy Dwelling. The goal of the Minimum Energy Dwelling is to reduce energy use 50% (by means of energy conservation features such as insulation, window-shading devices, color, ventilation, prevention of infiltration) and to obtain much of the remaining 50% from a 320 ft² evacuated tube solar collector with a 460-gallon insulated heat storage tank. Solar energy plus energy conservation is considered an effective combination for this dwelling and others.

evidence was growing that direct solar energy could be effectively used to supply some portion of heating and cooling needs. And funds were increasingly available from government and private sources to do something about it.

The sun's energy generosity is no longer being taken for granted, and more and more it seems certain to be less and less wasted. Examples of important and spreading uses include *solar air heaters* for agricultural drying. *Solar stills* are coming along as well to provide drinking water from saline or polluted waters in freshwater-deficient areas of the world. Major stills already are operating in the Australian Outback, Africa, the Middle East and southern Russia. The development of *rooftop solar stills* is recognized as a practical answer for places such as islands in the West Indies with rainfall and expensive importation by ship as their only sources of freshwater. In this simple but vital way the sun's energy can be used to make life easier and richer for many people.

Electric Power Generation

A key test of solar energy now and in the future will be its capacity to deliver electricity economically. Because our civilization is so totally dependent on electricity, from brushing its teeth to cooking its food, future energies no doubt will continue to be measured in terms of their ability to produce significant amounts of electricity. Whatever the energy source, efficient conversion to electricity will be essential. Fortunately there are a number of promising ways to transform solar energy into electricity. The solar-thermal system and the photovoltaic conversion system are two.

Solar-Thermal Generation of Electricity

The solar furnace, such as the ambitious one in the French Pyrenees (Figure 5), illustrates electrical generation from solar-thermal power. Another example is the solar farm depicted in Figure 13. The "farm" utilizes a large number of linear reflectors and absorber pipes that hold the radiated heat.

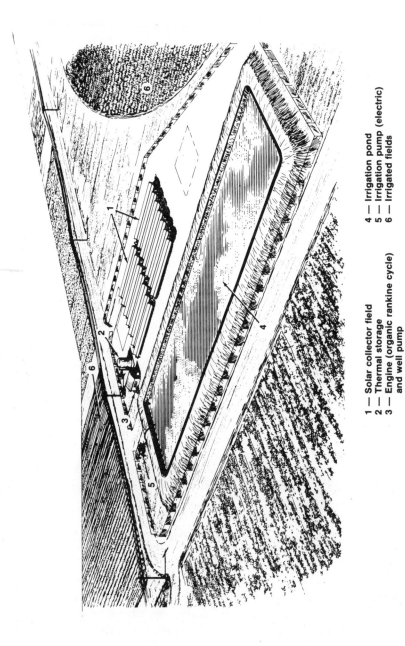

1 — Solar collector field
2 — Thermal storage
3 — Engine (organic rankine cycle)
     and well pump
4 — Irrigation pond
5 — Irrigation pump (electric)
6 — Irrigated fields

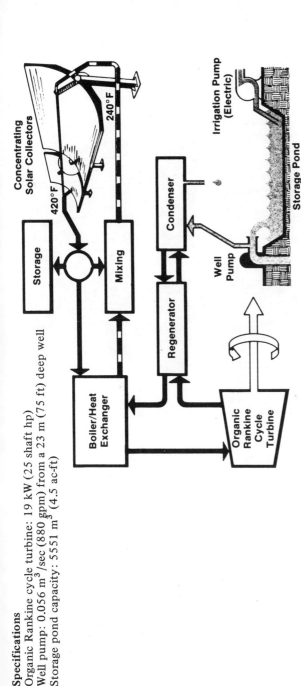

Specifications
Organic Rankine cycle turbine: 19 kW (25 shaft hp)
Well pump: 0.056 m³/sec (880 gpm) from a 23 m (75 ft) deep well
Storage pond capacity: 5551 m³ (4.5 ac-ft)

Concentrating Solar Collectors

420°F

240°F

Storage

Mixing

Boiler/Heat Exchanger

Regenerator

Condenser

Organic Rankine Cycle Turbine

Well Pump

Irrigation Pump (Electric)

Storage Pond

Figure 12. *Solar irrigation system for Southwest agriculture.* ERDA and the State of New Mexico are joint sponsors of an experiment in solar-powered irrigation conducted near Albuquerque. The ultimate purpose is to operate pumps by solar power for farm and ranch irrigation in the Southwest. Concentrating solar collectors pictured in this drawing from the Acurex Aerotherm company of Mountain View, California, are the heart of the system. The collectors heat an oil-like transfer fluid selected for its heat stability at high temperatures. Heat from this fluid changes Freon to gas, which drives the turbine to pump water into the plastic-lined storage pond.

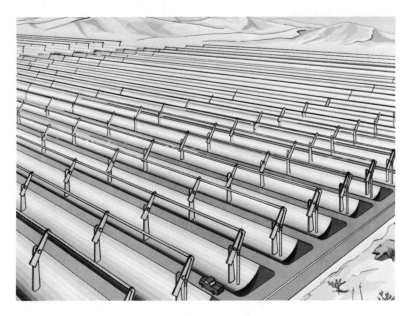

Figure 13. Artist's concept of fixed-mirror solar concentrators showing the mirrors and the tracking heat absorber pipes.

A variation on this solar farm approach involves the use of several thousand heliostats or clockwork-controlled tracking mirrors in a field. These mirrors follow the sun and reflect sunlight to the summit of a tower where heat is collected in a fluid such as water/steam or a eutectic salt. Enormous heat can be generated by this method for conversion to electricity.[6]

Using such solar-thermal installations to generate electricity on a large scale is a prime objective today. Since intense heat is a natural result of the solar-thermal process, it is also seen as a potentially practical way to supply industrial process heat. This could be an exceedingly valuable use of solar energy since approximately 30% of U.S. fuel consumption is for process heat at moderate or high temperatures. By establishing solar-thermal operations near industrial plants, the use of nonrenewable energies could be significantly reduced. The Sohio uranium mining/milling plant in Grants, New Mexico, will be the site of a solar-thermal facility designed to supply low-temperature process heat.[19]

Solar-thermal conversion is an important future energy technique that seems likely to be technologically refined and to grow, but it can hardly be called a new technique since the principles and essential methodology of solar-thermal conversion have been known almost a century. The use of sunlight to run a steam engine was demonstrated at a Paris exhibition in 1878, and sunlight was used in New Mexico in the 1920s by Harrington to generate electricity used to illuminate a mine.[20]

The point is that if we run seriously short of energy through inaction and failure to develop plentifully available energies, we can't blame technology for failing to deliver pragmatic solutions. Those who look to technology to save us when we start running out of gas, simply aren't paying attention. In solar energy, technology has already pointed out several viable directions.

The Photovoltaic Effect

Photovoltaic power generation, utilizing solar cells, is a more advanced means of electrical generation from solar energy, with the advantages of low maintenance requirements and impressive dependability. Solar cells are used in arrays and have proved highly successful in a number of limited projects (navigational lights, solar-powered buoys, etc.). In terms of future energy contributions, they have already justified technological optimism.

During space flights, a silicon solar cell has been used as the main source of electricity. The cell is based on the phenomenon known as the photovoltaic effect, or ability of light to effect an electrical flow in certain substances. This fact has been observed and studied since the nineteenth century, beginning with the work of Antoine César Becquerel, an electrochemical pioneer, who described the effect in 1839. It was 1954 before researchers at the Bell Telephone Laboratories and RCA produced the first silicon solar cells.

73

The core of a solar cell is an extremely thin slice from a silicon crystal containing a trace of boron (for positive electric charges) and a trace of phosphorus on top (for negative electric charges). Boron and phosphorus serve in much the same fashion as the positive and negative poles of a battery. Electric current moves in the cell when light rays strike the silicon crystal. The current can be drawn off as electricity. This was successfully done during space flights and on NASA's Skylab, which used a battery of solar cells for power throughout the 171-day flight.

The development of improved solar cells is naturally an active goal of private and governmental researchers. In the 1977 National Energy Plan, the photovoltaic effect receives the promise of both effort and money:

> Photovoltaic systems, using cells developed in the space program, are economic today for certain small, decentralized applications. These systems have a potential for dramatic price reductions that would make them economical for a broader range of applications. Increased funding is proposed to accelerate the development of economic photovoltaic systems.[7]

In 1977, cost is the main problem as with most other aspects of early solar energy use. We have the technology to obtain heat and cooling and electricity from sunlight, but the price is still too high. In the 1970s, a solar cell arrangement would cost $20,000 to $80,000 per peak kilowatt, compared with $500 per kilowatt for conventional steam electric generators using conventional fuels.[21]

But as the National Energy Plan suggested, new methods hopefully will bring remarkable savings in manufacturing solar cells. Research efforts and better production techniques are already reducing costs. Even before solar cell economics reach the $500/kW level, solar cells are finding practical applications to meet special needs as in the space program, electrical power on offshore oil rigs, and fast electrification programs in countries such as Iran, with an impatient yearning to eliminate delays in reaching modernization goals.

74

Prospects for technical breakthroughs in the production of solar cells seem better than ever. One approach, researched at the Mobil Tyco Solar Energy Corporation in Boston, is based on the use of a specially designed machine intended to produce continuous ribbons of silicon crystal, each several hundred feet long. Standardizing and expediting production of silicon crystal, main component of the solar cell, would solve a lot of problems toward making the photovoltaic effect energy-effective on a massive scale. The time might come, wrote John Wilhelm, when "solar cells are delivered to a house like rolls of roofing paper, tacked on, and plugged into the wiring, making the home its own power station."[11]

Silicon isn't the only semiconductor displaying the photovoltaic effect. Alternates such as cadmium sulfide-copper sulfide have been investigated. So far single-crystal silicon cells have a strong lead, but this may not continue indefinitely. Silicon is abundantly and inexpensively available. In the earth's crust, it is the second most plentiful material, 28% by weight, after oxygen, 47%. High-purity silicon is needed in transistors and, as we have seen, in solar cells. The complication has been the almost sixty-step, deucedly expensive method involved in obtaining suitable silicon crystals. First it is necessary to grow a single crystal from molten silicon using the Czochralski process and then repeatedly to polish it with great loss of material. But progress continues. Hopes are high. And it is unlikely that the photovoltaic possibility will be allowed to hide its magic permanently from mankind.

There are other examples of explorations underway and breakthroughs being achieved. More will come because the effort is being invested. Clearly the challenge of making solar energy competitive with other forms of energy has been taken up. It is probably only our imaginations if we seem to hear from far off a sound like solar thunder amiably saying, "About time."

Superconducting Solenoid

Because sunlight is intermittent rather than constant, storage is a critical challenge in solar-thermal systems, house heating/cooling applications, and most other uses needing around-the-clock energy supplies.

Solar homes have used various means of storing daytime heat for night use. The simplest answer may be that in Harry Thomason's first solar house, with tons of rocks in the basement surrounding the system's water tank. By absorbing heat during warm daylight periods, the rocks have it available for delivery during cooler periods. Variations on this basic method of storage embody much the same principle.

More sophisticated storage is needed, however, and a super-conducting solenoid device is envisioned as the answer for storage of electricity generated by solar energy. As in other solar energy challenges, further research is needed and is being carried out to achieve greater efficiencies and economies.

Irony may be detected by some in the fact that when considering solar energy we are struck by the ample availability of more energy than we know what to do with, and we immediately become bogged down in economics. It has often been observed with blithe irreverence that if all economists were placed end to end they wouldn't reach a conclusion. But this is an example of sacrificing fact for a good line. Economists do reach one clear and persistent conclusion: People normally choose the lowest possible price. This entrenched habit explains why users continue to use up dwindling supplies of oil and natural gas instead of rushing headlong and pell-mell for the nearest solar energy equipment store. Yet this same economic practice in time may well reverse the situation.

VARIATIONS ON A SOLAR THEME

Solar cities? Solar energy factories in space? These may be science fiction concepts now but wait. The future is on its way. We are moving forward to strange rendezvous in the decades ahead, and perhaps those decades are not as far ahead as we might imagine.

Solar Energy from Space Orbit

"Collection of solar energy by space satellites has been proposed, and the concept deserves further study," says the 1977 National Energy Plan.[7] This concept involves a solar array in geosynchronous orbit with the earth. The solar array and its concentrator would be located in space where solar energy could be transmitted to earth stations in the form of microwaves. According to estimates a power satellite of this sort would deliver more energy in its first year than was needed to produce it and put it in orbit.[22]

Energy from Manufacturing Facilities in Space

Conceive if you will a community of 10,000 people orbiting the earth. It is a Space Manufacturing Facility (SMF), made from moon materials. The SMF is powered by solar energy, and it exists to collect one product and dispatch it efficiently to earth. The product: energy. Construction and maintenance of Satellite Solar Power Stations (SSPS) is the function of the SMF. It converts solar energy to electricity and distributes energy to stations on earth by microwave transmission. The plan sounds Utopian, but the economics and logistics have been worked out by Princeton physicist Gerard K. O'Neill, and other scientists accept his conclusions. O'Neill considers present technology sufficient for the space community's construction and operation, and he believes 5000-megawatt SSPS's could deliver electricity to earth for less than .02 per kilowatt-hour, contrasting with .08 per kilowatt-hour paid in the mid-seventies by consumers in many American cities.

> An SSPS built at a space colony would be considerably simpler than one launched from the earth, because the colony-built SSPS could be designed without launch vehicle constraints. Turbogenerators could be fewer and of the most efficient size rather than kept within vehicle limits. Solar reflectors and waste-heat radiators could be built in large sizes and would never have to withstand launch accelerations. That is a significant advantage because an SSPS would be mechanically fragile.[23]

According to O'Neill, an SMF could be in space, on the job, in the 1990s. However logical and manageable the scheme might be in terms of scientific principle, it sounds too ambitious and grandiose—some might say wacky—to attract conservative investment dollars in the foreseeable future. But informed scientists seriously present it as one of our energy options for the future. Perhaps they are intrigued by the challenge or have been emboldened by television space odysseys. They have also performed the necessary calculations and determined that sky factories are theoretically possible. Nevertheless, there may be a considerable delay before such communities exist and U.S. postmen have to start regular mail deliveries. Ten thousand volunteers aren't likely to move aboard an SMF anytime soon. Some day? Well, men keep doing strange and amazing things.

The SSPS may be considerably closer. As currently planned, such satellites would be extended solar wings orbiting the earth at 22,300 miles and supplying up to 5,000 megawatts by microwave to an earth antenna 5 miles in diameter. This expansive, and initially expensive, energy apparatus is proposed by Dr. Peter Glaser, energy expert of the Arthur D. Little organization. Such a satellite would be a constant energy source, since it would always be in sunlight except during eclipses.

The O'Neill and Glaser schemes are bold, with science fiction echoes, yet boldness will be called for to develop and supply future energies on the scale demanded by the needs of civilization. An SMF with a docile herd of SSPS's around it is hard to see today. But tomorrow?

Solar Cities

A resort-retirement community in Arizona advertises itself as "Sun City." The community isn't operated by solar energy. It simply gets a lot of sunshine. Much of Arizona gets a lot of sunshine and has been an active solar research and testing area for many years. Today, even more than previously, Arizona looks to the sun. To encourage solar building, homeowners are allowed to deduct the cost of a solar installation from taxable income over a period of 5 years. Arizona talk about solar cities somehow doesn't sound like a pipedream.

Tomorrow, new cities will rise on the Arizona desert. These cities will be free from much of the pollution that afflicts us today, thanks to an endless supply of pure, clean energy from the sun.[24]

Strip mining of copper has left enormous craters in various parts of Arizona. Such craters are eyesores now and too large to fill. But even a big hole in the ground can be put to use. The idea is to use these craters as sites for vast fields of solar concentrators positioned to follow the sun. This idea too sounds like one whose time will come.

THE FUTURE?

When will this future energy be present energy? When will it contribute in a major way to the overall energy consumption pattern? As these questions are asked, energy economics promptly take over. Making solar energy competitive through advancing technology or through higher prices for other energies is the only answer available, and it is difficult to attach realistic time schedules.

The use of solar energy extensively for heating and cooling applications may be close. The trend seems to be in that direction, and in a few years, solar-energized structures may cease to be novelties. Homeowners already are being attracted by the dream of escaping natural gas and fuel oil bills. And solar energy proponents can paint a pretty picture of the satisfactions and peace of mind achieved through using an energy that is nonpolluting, almost trouble free, and reliable year after year. A solar system breakdown? "There's only one thing that can go wrong with it," said a salesman of solar energy collectors for the home, "and that's a leak." In future ambitious installations, potential difficulties may be more complicated; but trained technicians will be available to take care of them. Should a homeowner decide to use his solar system to generate electrical power, which is also feasible, he could free himself from the danger of power shortages and perhaps even electric bills.

The initial cost for an efficient heating and cooling system is high at the moment. Over a period of years, however, in some areas of the U.S. such systems pay for themselves. The payback period should decrease as costs decrease through a widening market and mass production. Even if solar heating and cooling aren't as cheap as other energies right away, some users will turn to solar anyway because of a desire to move in the direction of energy self-sufficiency. With a multifaceted solar setup, they can assert their independence of gas, oil and light companies. There is evidence that this independence has spreading appeal.

More ambitious use of solar energy to manufacture electricity, for instance, eventually will come, but probably not soon because of price. Solar energy could be used, despite its high cost, to save natural resources such as oil for future needs. Government legislation would probably be the only way to accomplish this, and no government is taking the lead aggressively in forcing the utilization of renewable energies to save depletable ones. The argument is even made that it pays to burn a barrel of oil now at $11 or $12 because it will cost much more to burn that same barrel in the future. Saving the barrel of oil for specialized future uses is a likable idea, of course, but it is one of those idealistic concepts that is singularly difficult to translate into action.

Fortunately, solar energy will still be available when man is ready, though it is too early to gauge the ultimate contribution of solar energy systems to man's energy needs. With new technology, some systems now little heralded might surge to the forefront. Others, for which liberating technologies fail to appear, might fade in promise and continue to contribute in only minor ways. There are those who believe that eventually *only* solar energy directly or indirectly from wind, oceans, bioconversion, etc., will be available to man.

Meanwhile the struggle to find practical ways of using the sun's energy continues. In some areas, it is intense; and the world, prodded by necessity, is coming awake to the realities of solar power.

In *Walden,* the American naturalist Thoreau wrote: "Only that day dawns to which we are awake. There is more day to dawn. The sun is but a morning star."

As we come awake, hopefully the habits of waste will be replaced by the instincts of conservation, and better use will be made of the steady dividends from our morning star. One truth is plain: The flood of solar energy fortunately will continue, and time will insist on its use.

Note: All solar references follow Chapter 5, Solar Energy and the Environment. See p. 89.

CHAPTER 5

SOLAR ENERGY AND THE ENVIRONMENT

SCRUTINIZING CLAIMS OF CLEANLINESS

In connection with the environment, an objective appraisal might conclude, accurately in both instances, that solar energy is both hero and villain. Solar energy is builder and architect of whatever environment exists. If yours is a halcyonic garden with green friends all about and a singing brook, give the credit to solar energy. If it is a lifeless desert with blistering heat, blame solar energy. If a slimy pond or polecat carcass turns fetid under the sun, solar energy has managed the unsavory trick. Without solar energy, pond or polecat would freeze solid and lie unfetid forever in eternal ice. Of course, without solar energy neither the pond nor the polecat would have existed in the first place.

Since solar energy is so inextricably a part of every environment, good or bad, almost by definition solar energy has to be viewed as one of the environmental constants and above reproach.

All right? Shall we agree solar energy is Miss Purity of all time and go on?

Yes, we can agree that solar energy is remarkably pure and nonpolluting in the standard meaning of the word, but instead of going on right away, maybe we should look at many of these assumptions. How often in past years has a new development been greeted as happiness and perfection incarnate, and we only learn some time later, too late, that it is wormy at the core and slowly afflicting everyone on earth with something unspeakable and incurable? If nothing else, the DDT experience *et al.,* has taught us wariness and prudent skepticism.

An author in *New Republic* stated the case for solar energy with a typical aria of devotion:

> It doesn't pollute or otherwise damage the environment. It creates no dangerous waste products such as plutonium. It won't run out for a billion years. It can't be embargoed by Arabs or anyone else. It's virtually inflation-proof once the basic set-up costs are met, and would wondrously improve our balance of payments. The technology involved, while still not perfected, is much less complex than nuclear technology . . . why then has it taken so long to discover the sun?[25]

We can't be entirely certain until solar energy is applied on a massive scale, but most of these claims seem to be essentially accurate. Solar energy doesn't add particulate matter to the atmosphere, but sunlight acting on debris scattered in the air by messy old mankind can produce temperature inversions and cause dangerous smog. Add sunlight to hydrocarbon and nitrous oxide automobile emissions and the result is a photochemical reaction with irritating and poisonous by-products. Sunlight in itself is the epitome of purity, perhaps, but sunlight plus industrial particulate effluents can set the stage for a rain of acids from the sky.

Solar energy doesn't contribute to the death of a river or the decay of an ocean. But the use of solar energy through ocean thermal conversion plants would inevitably affect ocean ecosystems; whether harmfully, beneficially, or both cannot be known fully until the plants are built. This is a familiar dilemma in assessing the environmental consequences of any project. Sometimes the consequences can't begin to be guessed with any accuracy until the project, for better or worse, is a *fait accompli.* This is why the argument that something shouldn't be tried until its full consequences are known is a rigorous formula for doing nothing. We have to assess and guess as well as possible, and then blunder ahead, making corrections as we can. Not to do something until we know how it will turn out is to say, "turn back, Columbus." But of course there is no turning back from the future. If we don't head for it, the future will come and get us. There is no doubt whatever that the human future includes solar energy.

Solar energy doesn't leave on our hands waste products that must be buried quickly and deeply for a quarter of a million years. The list continues.

There are many obvious environmental advantages in connection with solar energy, mostly advantages in fact. Nevertheless, the full exploitation of solar energy will have environmental effects that will not be exclusively positive and salutary. A solar plant capable of delivering several thousand megawatts of electricity will need as much as 40 square miles for monotonous arrays of solar collectors.[26] Most areas will not find such an appetite for territory easy, convenient or economical. Harvesting our gift from the sun is technologically demanding and physically difficult. The gift is lavish, but it doesn't arrive in a neat, compact package. It is almost as if someone gave us a box of chocolates by climbing a black walnut tree and tying each chocolate to different branches high above the ground, tempting us to break our adventurous necks as we collect the chocolates.

Another problem is that a solar facility large enough for major output might affect earth's surface reflection in a particular area and alter local weather for good or ill. There have been many proofs in recent years that environmental alterations of any kind always have an ill or two, whatever good results might accrue.

A large solar system at times would have surplus heat that would have to be discarded, just as waste heat must be handled from any other type of power plant.[27]

Other environmental drawbacks will appear as solar utilization expands. It would be naive to expect otherwise. The list of drawbacks may remain encouragingly small. Solar energy is not conveniently packaged for easy use, but it is essentially benign and impressively nonpolluting to the extent it is now known.

THE PUZZLE OF DELAY

Quantitatively solar energy dwarfs other energy forms and necessarily has from the start of life on earth. Qualitatively solar energy also appears to be unsurpassable. This superiority should continue indefinitely.

Sunlight has been coming to this planet a long time and has never changed its own character or quality. It has remained a fundamentally innocuous energy form, even if some of its offspring (the fossil fuels) have misbehaved environmentally. The parent is not responsible for the toxic wastes of its progeny.

Until recent years men used solar energy only for one purpose: to keep themselves alive. They didn't use it to run their machines, factories or floodlights. So environmental criticism of solar energy could have no validity in a practical sense, fetid ponds and polecats notwithstanding. If a standard of energy purity had to be selected, solar energy would be the judge, jury and victor.

Environmentalists readily accept that verdict. As almost evangelical critics of fossil fuels and fission power plants because they are "environmentally dirty," environmentalists are today's leading advocates of immediate solar energy use, particularly in heating and cooling where practical technology is available. A continuing and vital goal of the environmental movement is "pollution prevention technology." Under this classification would fall the technology needed to help solar energy societies replace fossil energy societies.

Such points were strongly made and goals stressed at the 1974 Air Pollution Control Association convention held in Denver, Colorado. Participants in a symposium concerning alternate energies predominantly supported solar energy as the preferred source for heating and cooling immediately, and for electrical power eventually when suitable technology is perfected.[28]

The environmentalists were displeased that large-scale research and development have not yet been financed and launched to solve the engineering challenges of solar energy. The fact that "only one penny is being spent on solar power for each 85 cents spent on nuclear development" received stern emphasis, as well as the fact that heating and cooling consume 25% of the energy used in the U.S. and so far no concerted effort is underway to guarantee accelerated use of solar energy for these functions.

Why the slowness? Suspicions were plentiful, and fingers were pointed accusingly at suppliers of traditional fuels who have economic reasons to keep solar energy under wraps, at governmental apathy, and at the public which has let itself be misled into believing that solar energy is a "sometime" possibility rather than a "right now" reality.

The cheerful greed of businesses for more profits, even if it means dismissing sunlight in favor of environmentally destructive oil shale development, and sleepy governmental apathy are old stories in the energy field. Perhaps governments, frightened by shortages, aren't as sleepy as they once were; but fresh drowsiness would no doubt quickly develop, given a modest interlude without crises.

The public too, of course, deserves a full share of environmental frowns. The public has guzzled energies and poured toxic wastes into the air it breathes as if there were no tomorrow worth concern. The public has and is using dirty energies and pleading for more. Curiously, or perhaps predictably, environmental appeals to neither conscience nor reason have been particularly effective in altering public attitudes or use habits.

The reason is not elusive. The public for decades has been encouraged by business, advertising and an indulgent way of life to consume rather than save, to ignore evidence of pollution, to pretend today's excesses will have no consequences tomorrow, and to assume those dirty but essential fossil energies will last one day longer than forever.

The evidence now is that nature has an effective way of calling a halt by turning off the tap. The dirty energies, it turns out, are not going to last forever and a day. The life-or-death question is just how sick the terrestrial environment—air, earth, oceans—may become before those energies are finally gone.

Some observers fear a salvageable environment may not survive the race. Others note the calamity of environmental degradation, but remain hopeful. Jacques Cousteau has written of his milieu, the sea, that "everywhere are sticky globs of oil, plastic refuse and unseen clouds of poisonous effluents. Is all now lost? I do not believe it." Cousteau's hope is that the oceans, and thus the world, will be saved for life because "the perceptive few who have the opportunity to see the ultimate disaster ahead" will join together "to warn the slumbering many."

Perhaps the many are no longer slumbering quite so soundly. Necessity is a determined alarm clock, and it is beginning, very reluctantly, to be heard. When nature says of the polluting energies, "that's all there is, there isn't anymore," comes the fateful dawn! What happens at dawn? Thoreau reminded us. Sunrise.

Senator Hubert Humphrey, a veteran advocate of solar energy, has expressed the wish that we could develop solar energy for the needs of all men. We should be able he said "to help other people who are desperately in need of energy resources and help them in a way that in no way threatens the peace of the world." Or, he might have added, the environment.

That help can still be provided. Perhaps in time it will be provided, with direct benefit at the same time to the air, the earth, the oceans. Luckily for us the sun shows no weariness with our delay and remains patient with its willful offspring.

A SELECTED SOLAR VOCABULARY

Absorptance - Soaking up heat in a solar collector, expressed as percent of total radiation available.

Bioconversion - Solar energy converted by plants and other life forms.

Biofuels - Energy from living things when converted to organic fuels.

British Thermal Unit (Btu) - Amount of energy needed to heat one pound of water one degree Fahrenheit.

Collector Tilt - Angle a solar collector is tilted to face the sun.

Concentrator - Apparatus designed to concentrate or focus sunshine into a small area for higher temperatures.

Emittance - Heat radiated away from the solar collector, expressed as a percent of the energy absorbed by the collector.

Flat-Plate Collector - Panel for converting sun radiation into heat, and transmitting the heat to a circulating fluid such as water or air.

Glauber's Salt - Inexpensive storage material for solar heat. Absorbs heat which can be released at night or during periods of cloud coverage.

Heliostat - Fixed mirror for reflecting sunlight into a solar collector or solar furnace.

Insolation - Amount of solar energy received per unit of horizontal area.

Kilowatt - One thousand watts of electricity, equals about 1-1/3 horsepower.

Langley - Unit measurement of insolation (one langley = one gram-calorie per square centimeter).

Megawatt - One million watts (about 1300 horsepower).

Noon Mark - Mark used by early settlers and farmers on the south side of a house. Shadow falling from a vertical object indicated Noon.

Ocean Thermal Energy - Solar energy in the water of the ocean surface.

Photolysis - Chemical decomposition brought about by radiation, such as solar radiation.

Photosynthesis - Chemical compounds synthesized as a result of solar radiation. Formation of carbohydrates in chlorophyll tissues of plants.

Photovoltaic - Electricity produced by action of sunlight on a solar cell.

Pyroheliometer - Measuring device for solar radiation.

Quad - Quadrillion Btu (U.S. energy needs might equal 200 Quads by year 2000.)

Rock Storage - Container for rocks used to store heat from solar collectors (about 50 lb of rocks needed for 1 sq ft of solar collector area).

Selective Surface - Coating for solar flat-plate collectors that greatly reduces emittance by absorbing most of solar radiation.

Solar Cell - Instrument, usually containing silicon wafer, for converting sunlight into electric current.

Solar Constant - Average amount of solar radiation reaching the earth's atmosphere per minute, slightly less than 2 langleys.

Solar Cooker - Device for cooking with solar energy.

Solar Furnace - Device for focusing solar energy to achieve high temperatures, up to 2000° Fahrenheit and higher (record for a solar furnace is 5600°F).

Solar Power Farm - Extensive assemblage of solar collectors or concentrators to generate electricity with solar energy.

Solar Still - Installation for use of solar energy in desalinization of water.

Sun - Earth's life source. Thoreau called it our "morning star." Da Vinci said it heats the universe. We couldn't live without it. Our special fusion friend.

Sun Tracking - Movement of a solar collector to follow the sun.

Trickling Water Collector - Solar collector designed by Dr. Harry Thomason. Utilizes a trickle of pumped water down the valleys of a corrugated surface.

Wind - One of many ways solar energy chooses to express its emotions. Can be put to work by man.

SOLAR ENERGY REFERENCES

1. Gough, W. C. "Fusion Energy and the Future," *The Chemistry of Fusion Technology,* D. M. Gruen, Ed. (New York: Plenum Publishing Corporation, 1972).
2. Gough, W. C. and B. J. Eastlund. "The Prospects of Fusion Power," *Scientific American,* February 1971, pp. 50-64.
3. Hammond, Allen L. "Solar Variability: Is the Sun an Inconstant Star?" *Science,* Vol. 191, No. 4232, March 19, 1976, pp. 1159-1160.
4. Williams, J. R. *Solar Energy—Technology and Applications,* Revised Edition (Ann Arbor, Michigan: Ann Arbor Science Publishers, 1977).
5. Hubbert, M. K. "Industrial Energy Resources," *Nuclear Power and the Public,* H. Foreman, Ed., (Minneapolis, Minnesota: University of Minnesota Press, 1970), pp. 179-206.

6. Morse, Frederick H. and Melvin K. Simmons. "Solar Energy," *Annual Review of Energy*, Volume 1, 1976, (Palo Alto, California: Annual Reviews Inc., 1976), pp. 131-158.

7. The National Energy Plan, Executive Office of the President, Energy Policy and Planning, April 29, 1977, pp. 75-77.

8. *The New York Times*, July 17, 1977, Section 4, p. 1.

9. Gordon, Howard and Roy Meador. *Perspectives on the Energy Crisis*, Volume 1, (Ann Arbor, Michigan: Ann Arbor Science Publishers, 1977), p. 5.

10. *Solar Energy Intelligence Report*, Volume 3, No. 20, June 27, 1977, p. 134.

11. Wilhelm, John L. "Solar Energy, the Ultimate Powerhouse," *National Geographic*, Volume 149, No. 3, March 1976.

12. Solar Supplement, *The National Observer*, May 2, 1977, p. 1.

13. Edmondson, W. B., Ed. "Breakthrough in Selective Coatings," *Solar Energy Digest*, Volume 2, No. 2, February 1974, pp. 1-2.

14. *Moscow News*, "Solar Energy," No. 2, January 1974.

15. Sargent, S. L., Ed. *Proc. Solar Collector Workshop*, Rep. No. NSF/RA/N/75/015, National Science Foundation, Washington, D.C., May 1975.

16. DeWinter, F., Ed. *Proc. Solar Cooling Workshop, Los Angeles, February 6-8, 1974*, Rep. No. NSF/RA/N/74/063, National Science Foundation, Washington, D.C., 1974.

17. Levy, Lawrence G. "Sun Power Reaching the Age of Application," *The New York Times*, September 21, 1975, p. F4.

18. Duffie, John A. and William A. Beckman. "Solar Heating and Cooling," *Science*, Vol. 191, No. 4223, January 16, 1976, p. 149.

19. Lawrence Livermore Laboratory, Solar Energy Group. *LLL-Sohio Solar Process Heat Project, Report No. 2, 1 May 1974*, Rep. No. UCID-16630-2, Livermore, California, 1975.

20. Daniels, F. *Direct Use of the Sun's Energy* (New Haven, Connecticut: Yale University Press, 1964).

21. Claassen, Richard S. "Materials for Advanced Energy Technologies," *Science*, Vol. 191, No. 4227, February 20, 1976, p. 743.

22. Glaser, P. E., "Solar Power Via Satellite," *Astronautics and Aeronautics*, August 1973, pp. 60-68.

23. O'Neill, Gerard K. "Space Colonies and Energy Supply to the Earth," *Science*, Vol. 190, No. 4218, December 5, 1975, p. 945.

24. *Arizona Highways*, Vol. LI, No. 8, August 1975.

25. Barnes, Peter, "Who'll Control Sun Power? The Solar Derby," *New Republic*, Vol. 172, February 1, 1975, pp. 17-19.

26. Clark, Wilson. *Energy for Survival*, (Garden City, New York: Anchor Books, 1974).

27. Carter, Luther J. "Solar and Geothermal Energy: New Competition for the Atom," *Science*, Vol. 186, No. 4166, November 29, 1974, p. 812.

28. *The Denver Post*, "Environmentalist's View," June 12, 1974, p. 36.

CHAPTER 6

COAL AND HYDROGEN:
TWO OLDTIMERS WITH A FUTURE

"The mine is dark . . . . If a light come in the mine . . . the rivers in the mine will run fast with the voice of many women; the walls will fall in, and it will be the end of the world . . . So the mine is dark . . . But when I walk through the shaft, in the dark, I can touch with my hands the leaves on the trees, and underneath . . . where the corn is green . . . There is a wind in the shaft, not carbon monoxide they talk about, it smell like the sea, only like as if the sea had fresh flowers lying about . . . and that is my holiday."

Miner Morgan Evans in "The Corn Is Green"
by Emlyn Williams

Estimates vary. Some say 50 years, some say 25, some say less. But nearly all energy projections show natural gas running out soon and then oil, at least in their ability to supply the bulk of our energy needs. Ironically, when oil, which replaced coal as the number one fuel, assumes a secondary role again, it in turn may for a time be replaced by old king coal again.

THE OLDEST NEW ENERGY IN TOWN

Now we think of coal as a black, rock-like substance largely composed of carbon that can be used in various ways as a fuel, some of them environmentally harmful. It is strange but true to reflect that coal began as plants growing lushly in swamplands some three hundred million years ago in the Carboniferous

Period. Now living things from that remote time can be burned for heat and light. We burn the past to warm and illuminate the future. Except for wood, no conventional fuel that has been a major contributor to man's power consumption record has a longer history than coal. The *Old Testament* prophet Isaiah describes a six-winged seraphim with "a live coal in his hand." Coal has been mined in Europe since the Middle Ages. Written references to coal in English abound since the ninth century. Many fields still supplying coal in modern times were originally developed as early as the 1200s. Underground coal mines were already known in medieval times, and atmospheric pollution from coal dates back at least to 1257 when Queen Eleanor of England was forced to leave Nottingham Castle because of coal fumes. Another English Queen, the first Elizabeth, was also reported to "Findeth hersealfe greatly greved and annoyed with the taste and smoke of the sea-cooles."[1]

In the eighteenth century, Benjamin Franklin was in England on business for the American colonies. His wife wrote that he should burn wood instead of coal so the air in his house would be purer. Franklin replied burning wood was a waste of time unless everyone did. "The whole town is one great smoaky house, and every street a chimney, the air full of floating sea coal soot."

Despite the long use of coal by man, in the last quarter of the twentieth century it is still astonishingly plentiful in many areas. The U.S. has an estimated 4 trillion tons (90% of U.S. conventional energy reserves).[2]

Why is so much coal still available despite its extended history? The answer is simple. There was an enormous amount of coal to start with when men discovered that though it burned smokily, much of it, and irritated Queens, nevertheless it did burn and substituted quite ably for wood when necessary. So coal was used and in large amounts for a considerable time, but this use in a sense was merely scratching at the surface of the gigantic coal reserves still waiting in the earth.

Another factor explains the abundance of coal in the twentieth century, the most intense energy consumption era in the history of man. Because of its untidy nature, coal has generally been a last-resort fuel. When there was insufficient wood available, coal was used. When it was possible to switch from coal to something different, the switch was made. That is why we are running out of oil instead of coal. When oil became available, we used it instead of coal.

For example, railroads during their first century chiefly used coal as fuel in steam locomotives by necessity rather than preference. After World War II, railroads switched from coal-fired steam locomotives to oil-burning diesels. When that shift occurred, the use of gas, oil or electricity as home furnace fuels rather than traditional coal was a widespread and growing energy fact of life. The Clean Air Act of 1970 resulted in standards that prohibited certain harmful emissions such as oxides of sulfur which reduced still more the amount of coal used by industries and by power plants for electricity.[3]

The result is that plenty of coal is still available, because it wasn't used. Indeed, it was treated as an outcast when cleaner fuels could be found. Now that those cleaner fuels are becoming harder and harder to find, a telegram is being sent to the outcast: "Old King Coal, all is forgiven. Please come home."

The use of coal is now being encouraged by the U.S. government with anxious vigor. There is simply no other energy immediately available to replace oil.

> Expansion of U.S. coal production and use is essential if the nation is to maintain economic growth, reduce oil imports, and have adequate supplies of natural gas for residential use.[2]

The objective for coal in the future is easy enough to state in words: Rapidly expand the use of coal, accomplish this without the pollution associated with coal in the past, and find new technologies that will increase the versatility and convenience of coal in a broad spectrum of energy applications. The goal is easy to say, but implementation is another matter.

THE PROBLEMS OF COAL

No doubt one reason we still have plenty of coal is the fact that it has always been a problem-energy in terms of acquisition, transport and use. There is simply little that is convenient about coal except its availability.

Problems start at the mine. Coal is removed from the earth either through underground mines or strip mines. The men who work in underground mines know theirs is the most dangerous undertaking in American industry. Coal mining is estimated to be four times as hazardous as a typical manufacturing occupation. In the case of strip mines, the work is less hazardous for the men, but more hazardous for the earth. Strip mining can leave the earth with a gaping man-made wound, weakening the earth's natural defenses against erosion, landslides and floods. New federal regulations in the U.S. are seeking to set standards for strip mining to prevent permanent destruction of the land, but complying with these regulations will greatly add to mining costs.

When mined, transport is the next problem for coal, getting it to the point of use. Railroads are the traditional way to move coal in the U.S., but efficient transport of the gigantic quantities expected to be used will require major capital investments in railroad cars, improved roadbeds, and a general refurbishment of America's railroad system. An alternate way of moving coal in large quantities is to construct coal-slurry pipelines that would carry crushed coal mixed with water to the point of use. Slurry pipelines could be useful in conveying raw coal from mine sites in western states, but there are difficulties with the solution. Large amounts of water are required, and water is often the missing ingredient in any technological equation for western America. Also, effective coal-crushing technology has not been perfected, though the problem is being studied.

When coal reaches its destination, the challenge of using it without polluting the environment takes over. Even in the case of low-sulfur coal from western states as well as the high-sulfur coal from several states east of the Mississippi River, environmental protection safeguards are necessary. Environmental restrictions on the burning of coal raise costs, and they also raise the resistance of businessmen asked to convert their plants from oil and gas to coal. From 1975 to 1977, the U.S. Federal Energy Administration directed approximately 100 plants to switch, but less than 20% did so. Many of the others chose instead to resist through time-consuming appeals. In U.S.

industry there is clearly no eager rush to attack the coal obstacle course voluntarily or any sooner than is obligatory. Experts who view increased use of coal as essential, expect new technology to reduce some headaches of conversion.

ADVANCING COAL TECHNOLOGY

After supplying man with energy for at least 1000 years, coal in the 1970s should be surprised and flattered to find itself the focus of such intensive research. Work is going forward in both private and government-financed programs to find better ways of mining, transporting and burning "old filthy" as coal has been known in the past. The hoped for goal of all these programs is a rendezvous time somewhere in the future that will make coal fully and safely usable to produce process heat, electricity, and even liquid fuels or gas. There are few limits to the technological plans for coal, and the reason is uncomplicated: Coal exists. Energy is needed, and the need may become desperate. So let's do something about coal. Something is being done and on a growing scale with a number of highly promising technological pathways to explore.

Scrubbers and Other Coal-Cleaning Efforts

Mechanical scrubbers are the simplest approach to prevent sulfur dioxide contamination of the atmosphere as a result of burning coal. Effective scrubbers exist, but their installation and operation are costly and troublesome. Thus, finding more efficient ways to clean flue gases is a major search effort. Flue gas desulfurization (FGD) scrubber systems are already in use. New and better systems are being developed.

One promising new direction is "cleaning" coal before it is burned and thus eliminating the danger of sulfur emissions in advance. Using technology pioneered in Germany, Stephen Krajcovic-Ilok has developed a coal reduction process to grind coal into 4-micron size particles that are almost pure carbon, with ash pyrite and sulfur easy to remove and industrial scrubbers unnecessary.[4]

95

Another approach is using solvent-refined coal processes to remove sulfur content chemically. A demonstration plant for this method is on the government list of research programs. In a TVA news release for June 10, 1977, announcement was made of plans for a coal-washing plant, the largest so far built in the U.S., to meet sulfur dioxide emission limits at the Paradise Steam Plant, Drakesboro, Kentucky. The plant will clean 2000 tons of coal per hour as it lowers sulfur content to meet the sulfur dioxide limit of 5.2 pounds per million Btu established for the Paradise Steam Plant. In 1976 the plant produced 11 billion kilowatt-hours at a cost of about $82 million. Meeting sulfur dioxide limits is estimated to add a cost of approximately $35 million. In the plant itself, the TVA intends to use large electrostatic precipitators to achieve better control of fly ash emissions.

Coal Gasification

In addition to cleaning coal for conventional uses, transforming it into gases or liquids for more convenient use is receiving close research attention and major investments. Through coal gasification, high-Btu fuels that may substitute for methane or natural gas are believed possible just over the technological horizon. Closer is low-Btu gas from coal gasification suitable for industrial applications. Coal gasification will make available vast underground reserves of currently inaccessible coal. One method of *in situ* gasification planned for implementation at western coal sites involves drilling two steel-lined wells, 60 to 120 feet apart, to the bottom of a coal seam. A heater is lowered down the "ignition well" and the coal lighted. Air is forced down the second well, and the air works its way through the porous coal seams (lignite and bituminous coal are the types that lend themselves to this method), while the fire works toward the air. When air and fire meet and a channel exists between the two wells, gasification starts. Researchers think *in situ* gasification may supply 5-10% of U.S. energy needs in time, and according to ERDA estimates, 1.2 trillion tons of U.S. coal could be used this way.

The problem of land slippage as a result of the underground burning is solved by leaving unburned coal pillars for support.

Fluidized bed combustion of coal is a variation on low-Btu gasification that has proved effective in reducing sulfur dioxide emissions 90% or more. It works this way: Small particles of limestone and coal are suspended by a stream of air in a fluidized bed combustor. Heat from the fluid-like suspension can be used to produce steam which in turn can produce electricity in a conventional power plant. The limestone particles react with the sulfur dioxide present to form inert calcium sulfate. The more limestone used, the greater the reduction of sulfur dioxide that otherwise would add to atmospheric pollution.

ERDA estimates indicate that 60% of the sulfur dioxide annually discharged atmospherically in the U.S. comes from coal used to generate electricity. Fluidized bed combustion is seen as a practical means of meeting atmospheric standards without denying ourselves the energy of coal. This approach to burning our coal and keeping our health is believed both technologically and economically feasible now with commercial applications coming soon.

In May 1977 a $14 million power plant at Georgetown University was announced[5] in which high-sulfur coal that has been crushed will be forced through a heated bed of pulverized limestone. Limestone will ignite the coal and release heat for power uses. Simultaneously, the limestone will trap the coal's sulfur content, producing a residue that can be converted to gypsum and used in a variety of ways (*e.g.*, soil conditioning, on road beds, to make wall board).

These and many parallel efforts are confronting the problem of pollution directly. Two assumptions appear strong: that neither public nor government acceptance of environmental deterioration through coal pollution will any longer be granted and that technology can perfect ways to burn coal effectively without intolerable emissions into the air.

Coal Liquefaction

Researchers are busy and hopes are high that plentiful coal will eventually be transformed into liquid energy products that can be poured in the gasoline tanks of automobiles and the fuel tanks serving home heating furnaces. This is the number one dream for coal. If that 4 trillion plus tons can be partially used to produce gasoline and fuel oil substitutes, the future will seem considerably less static and chilly.

Converting coal to liquid fuels can already be accomplished— at a price. Germany during World War II obtained fuel in this manner using hydrogenation methods. A number of pilot plants in the U.S. are engaged in coal liquefaction and developing the process technology required for large-scale operations on a commercially viable basis. These developmental efforts have established the technical parameters of coal liquefaction but haven't met the engineering challenges for commercialization. The early 1980s are now seen as the time for testing and proving coal liquefaction techniques can function effectively when scaled up to full-size plants, producing as much as 100,000 barrels of oil daily from 30,000 tons of coal.

Coal liquefaction is a complex chemical process based on the interaction of hydrogen and a slurry of coal particles. Ideally, the liquefaction process itself will supply hydrogen through the formation of gaseous hydrocarbons or carbonaceous residues from coal. To be economical, the process should generate its own hydrogen for the continuing hydrogenation process.

The resulting fuel, depending on the amount of hydrogen used, "can range from light distillates, equivalent to refined kerosine and diesel oils suitable for gas turbine fuel, to a solid material that melts at about 400°F and is essentially free of ash and reduced in sulfur so that it is acceptable for combustion as a solid in electrical generation equipment of conventional design."[6]

Much work is ahead before coal liquefaction is a pragmatic reality, but the work isn't being neglected. Government funding for a 600-ton/day coal-to-oil pilot plant in Kentucky has been provided, and it has been emphasized that "the new coal technologies are critical to the National Energy Plan, both as an immediate aid in converting from scarce to abundant resources and as a future source of synthetic oil and gas."[2]

Natural Gas From Coal and Shale

Substantial deposits of methane are known to be trapped in coal deposits. Historically, methane has been the worst danger and source of fatal accidents in coal mines. Coal mine explosions have frequently been traced to pockets of volatile methane. Venting off methane from coal mines is a method of obtaining methane for use as natural gas and at the same time making mines safer. U.S. coal deposits are reported to contain 8.5 trillion cubic meters of methane and possibly much more.

Natural gas is also available from the Devonian shale deposits found in the U.S. (up to 650,000 square kilometers).

> This shale, which is more similar to coal than to western oil shale, contains about 0.63 to 0.95 cubic meters of trapped natural gas per metric ton. The USGS estimates that there are about 14 trillion cubic meters of gas in Devonian shale.[7]

COAL, COAL EVERYWHERE

The U.S. is so amply endowed with coal, using it for energy becomes almost a moral and practical obligation. Of the estimated 4 trillion ton reserves, current technology can mine approximately 500 million tons. The remaining 3.5 trillion tons are found in deep seams or thin ones that inspire present speculations and challenge future ingenuity. About one-third of this mammoth reserve is believed recoverable via underground coal gasification.

Given all of this coal, what does the U.S. plan to do about it? Are the plans energy-in-the-sky schemes, or can they be achieved?

With coal production in the second half of the 1970s at about 600 million annual tons, doubling this output by 1985 is the first goal; and by the year 2000, annual coal production of 2 billion tons per year is projected! According to Robert C. Cowen in *The Christian Science Monitor,* this "would mean investing over $100 billion (in 1975 dollars) for new mines, transportation, and other facilities, to say nothing about finding skilled manpower and coping with the environmental problems involved."[8]   An equally grandiose ambition foresees 5 million barrels of synthetic fuels from coal by 1995, a volume requiring 200 plants costing about $1 billion each.

Yet it is easy to look at numbers and be numbed. Whatever the scope of the challenge and the cost, it is probable that men will keep hacking away at their coal problems until they solve them one way or another. If not this year, next year. If not this millennium, next millennium. Coal traditionally shows remarkable patience with man's fitful exploitation. It stays put. It doesn't go away.

Among the unanswered questions are these: Will coal be taken from the ground or converted in place without wantonly assaulting the land? Will coal be transformed to electricity or heat or other usable energy without degrading the quality of the air we breathe? Will coal be intelligently developed and used as an interim energy while renewable energies are technologically refined, or will we become hooked on coal as we were on oil, assuming an eternity of the stuff and postponing the development of other energies? These are critical questions troubling the present, alerting the future.

UBIQUITOUS HYDROGEN

"Ubiquitous" is a useful word meaning "present everywhere, omnipresent." We don't find hydrogen everywhere, of course, but have you noticed how we keep encountering this remarkable element again and again when considering various forms of energy? Hydrogen is the basis of solar energy through thermonuclear fusion in the sun. Hydrogen or deuterium (an isotope of hydrogen) is a critical part of the fusion cycle in all the fuels being considered by nuclear engineers and physicists for nuclear fusion. The use of hydrogen in the hydrogenation of coal to produce liquid fuels has been suggested. Hydrogen is present in all the hydrocarbon fuels from methane (CH_4) to the complex mixtures of hydrocarbons, including gasoline and petroleum. Why do the gaseous and liquid hydrocarbons function so well as fuels—because hydrogen burns with such efficiency and enthusiasm whenever it gets a chance.

100

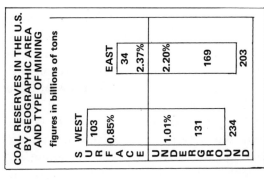

COAL RESERVES IN THE U.S. BY GEOGRAPHIC AREA AND TYPE OF MINING		
figures in billions of tons		
	WEST	EAST
SURFACE	103	34
	0.85%	2.37%
UNDERGROUND	1.01%	2.20%
	131	169
	234	203

percentage indicates sulfur content

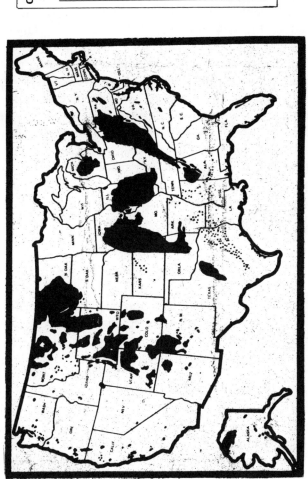

Figure 14. Areas of the U.S. with known coal reserves.

101

Hydrogen is an essential link in so many of our energy systems, ways to use hydrogen as a fuel are naturally being explored. If hydrogen is the heart of other energies, why not go directly to the heart for the cleanest and most effective energy fuel we know? There is, after all, more hydrogen in the universe than any other element (about three-fourths of the universe by mass, about 90% of the molecules).

Hydrogen can be obtained from water (H_2O) through electrolysis, from the action of mineral acids on metals, from hydrocarbons and hot coke through the catalytic reaction of steam. Each day, millions of cubic feet of hydrogen are manufactured in the U.S. and handled with care. "Handling with care" is indispensable with hydrogen. It was hydrogen that filled the ill-fated dirigible Hindenburg when it burned in seconds while mooring at Lakehurst, New Jersey, in 1937.

Developing hydrogen energy systems that are safe and efficacious is a key activity of contemporary energy research. Obtaining hydrogen thermochemically from water without the need for fossil fuels would be a major accomplishment. If such a system could be perfected and made economical, an energy option would be available from a virtually unlimited source. Progress has been made, but the technology for such a system still has a long journey ahead.

Hydrogen from nonfossil sources is an ultimate goal. Easier and faster to accomplish would be obtaining hydrogen from coal on a larger scale than in the past using coal gasification processes.

Paralleling the challenge of obtaining hydrogen economically from water or hydrocarbons are the manifold challenges of hydrogen transport, storage and use. Transport is expected to be solved via a network of steel pipelines. Storage possibilities give hydrogen advantages over difficult-to-store electricity, and make it a useful partner in a hydrogen-electric energy system, hypothesized as a likely application of hydrogen in the future. Hydrogen is more difficult to store than less unstable fuels such as oil, but it can be effectively stored several ways (*e.g.,* in underground chambers such as exhausted gas or oil fields, as a compressed gas for limited uses, in line pack storage similar to natural gas, at low temperatures as a cryogenic liquid in insulated tanks, or as metal hydrides which give up their hydrogen when heated).[9] Give your imagination free rein when it comes to the potential use of hydrogen fuel, including most of the ways we consume fuel individually today. Aircraft, automobiles, industry and homes have all been studied as practical energy consumers that could be technologically designed or adapted for direct use of gaseous or liquid hydrogen. The use of hydrogen for generation of electricity may ultimately prove the most beneficial application.

As with other energy sources, making hydrogen economical as a fuel or a reliable catalyst for alternate energy systems is the side of the court where the ball now is. In a universe filled with hydrogen, sooner or later we shall have to move for our energy in directions prescribed by the character and chemistry of this ubiquitous phenomenon. If we follow fusion or solar energy to a hoped for oasis of energy plenty, we accept hydrogen's lead whether we acknowledge it or not.

COAL AND HYDROGEN - REFERENCES

1. Galloway, R. L. *A History of Coal Mining in Great Britain,* Newton Abbot, David and Charles Reprints, 1969, p. 24.
2. The National Energy Plan, Executive Office of the President, Energy Policy and Planning, April 29, 1977, pp. 63-68.
3. *Exploring Energy Choices,* Preliminary Report of the Ford Foundation's Energy Policy Project, 1974, p. 14.
4. *Iowa Energy Bulletin,* Iowa Energy Policy Council, Vol. 3, No. 3, April 1977, p. 6.
5. *Environmental Impact News,* Vol. 3, No. 5, May 1977, p. 8.
6. Alpert, S. B. and R. M. Lundberg. "Clean Liquids and Gaseous Fuels from Coal for Electric Power," *Annual Review of Energy,* Volume 1 (Palo Alto, California: Annual Reviews Inc., 1976), pp. 87-99.
7. Maugh, Thomas H. II, "Natural Gas: United States Has It if the Price is Right," *Science,* Vol. 191, No. 4227, February 13, 1976, p. 549.
8. Cowen, Robert C., "Will U.S. Share its Coal?" *The Christian Science Monitor,* June 1, 1977, p. 29.
9. Gregory, D. P. and J. B. Pangborn. "Hydrogen Energy," *Annual Review of Energy,* Volume 1 (Palo Alto, California: Annual Reviews Inc., 1976), pp. 279-310.

103

CHAPTER 7

NUCLEAR FISSION

> "Surrendering to 'what if' is the death of science.
> When we allow 'what if' to dictate the future, there
> is no progress."
>
> Professor Terry Kammash
> Nuclear Engineering
> The University of Michigan

A NEW KIND OF FUEL

In 1945 when the atomic bomb helped bring an end to World War II and usher in the atomic age, poet Robert Frost wrote to his daughter about this event and its implications as he saw them:

> With things like the atomic bomb I pride myself on putting them in their relative place a priori without ever having been in a physics laboratory. I said right off uranium was only a new kind of fuel . . . . The mind keeps penetrating deeper into matter. I knew at once that the important question was how much uranium there was in the world . . . . The new explosive can be bad for us. But it can't get rid of the human race for there would always be left, after the last bomb, the people who fired it—enough for seed and probably with the same old incentive to sow it. There's a lot of fun in such considerations.

With the passage of more than three decades, the promise and the threat of nuclear fission seem to have brought the human race less fun than confusion, fewer answers than questions, and continuing controversy. In 1977, the likelihood that nuclear energy from fission reactors will contribute in a major way to U.S. energy still exists, but the prospects are dimmer because of breakdowns, unproven safety, and tireless criticism by numerous scientists and environmentalists.

The poet gave his daughter the facts of the matter when he described uranium as a new kind of fuel and argued that the important question is how much there is. Over three decades later, how much nuclear fuel in the form of natural uranium the U.S. can count on is still unknown precisely. ERDA through its National Uranium Resource Evaluation Program (NURE) has reported 1.8 million tons of uranium available, sufficient to fuel 350 to 390 nuclear reactors for estimated 30-year life spans. ERDA previously predicted 300 to 400 nuclear reactors would be constructed in the U.S. by the year 2000.

A study panel from the National Academy of Sciences challenges ERDA's 1.8 million ton figure as too high, and holds that no more than 1.76 million tons could be produced before the end of the century. The possibility of serious shortages after 2000 raises questions whether or not nuclear energy can become our number one energy supplier as proponents claim even if problems of safety and waste disposal can be solved. But there is clearly a lack of firm data available on uranium reserves. Reports from the Ford Foundation and the Pan Heuristics Institute indicate the availability of much larger reserves (from unexplored areas such as Alaska) sufficient for future needs.

HOW WE ARRIVED WHERE WE ARE

No one knows for certain how much uranium there is. No one knows for certain whether much nuclear fuel will ever be needed as the 1970s continue to be times of trouble for the nuclear industry. Construction has been stopped on some reactors. New reactors are not being ordered by electric utilities at the expected pace. The Nuclear Regulatory Commission in 1977 reported that if the trend persists there will be 106 nuclear reactors operating in the U.S. at the end of 1985, considerably fewer than the 200 ERDA forecast earlier.

What has gone awry with the nuclear promise? What are the chances it will go right again?

In the 1970s it has become clear that too much was expected from nuclear energy too soon. As with any major new energy form, patience is needed and sufficient time for suitable technology to be designed, tested and perfected at a measured rather than a hectic pace. Those who want to turn the atom fully on overnight are too hasty. Those who want to turn it entirely off permanently are too timid.

In 1977 there is no reason to assume that nuclear energy based on the fission process in nuclear reactors will cease to be developed. It is reasonable to expect that it will move forward much more slowly than originally hoped. The nuclear industry is harassed by a pack of difficulties from obtaining financing to keeping completed plants operating.

California, for example, has been reported to need 30 atomic energy plants by the 1990s, yet only 2 operate in the state today. One plant previously in operation at Humboldt Bay has been inoperative for a year, and the Nuclear Regulatory Commission is recommending that it be permanently inoperative because of earthquake danger. Another California plant, a billion dollar investment near San Luis Obispo, is also inoperative for the same reason.

The future of nuclear energy has become a scientific and political candidate for a crystal ball as the struggle to solve problems of safety, efficiency and financing continue. In 1975 President Ford recommended 135 new plants by 1985. This would require an average of more than a plant a month, yet in 1976 only three new plants were ordered, and none in 1977.

For several years a number of protest groups have opposed the construction of nuclear facilities until the hazards have been eliminated, but it is felt that the principal difficulties of the industry in the 1970s come less from the protesters than from operational problems within the industry itself. The majority of Americans have demonstrated in elections and opinion polls that they are in favor of nuclear energy, but the slowdown in new starts as well as the breakdown of existing plants has occurred despite the traditional vote of confidence by the general public.

Meanwhile critics have continued their antinuclear campaigns, dramatizing the potential hazards, the accidents, and the natural connection between atomic power plants and atomic bombs. There are obviously several reasons for the holding pattern on atomic energy, and since all cannot be resolved simultaneously, the command for full speed ahead is not anticipated soon.

Yet significantly, the "all stop" signal has not been given, nor is it expected. In 1977, 63 operating nuclear power plants supply approximately 10% of U.S. electricity needs. The President's National Energy Plan projects 75 additional plants and 20% of U.S. electricity from nuclear energy by 1985.[1] Simultaneously, the U.S. is deemphasizing work on the plutonium breeder reactor to minimize the danger of nuclear proliferation, and making available U.S. uranium enrichment services on a wider basis to other countries. Elsewhere in the world the Nuclear Rush is in no way abated.

During May 1977 France announced agreement to build two nuclear power generators for Iran. Cost of the two 900,000-kilowatt plants is about $2 billion. Iran was also reported to be purchasing eight nuclear plants from the U.S.—cost, $16 billion.

Other countries, especially those dependent on expensive imported oil, are committing themselves to nuclear energy on a large scale and refusing to let safety concerns stop them. By 1985 Italy intends to obtain 40% of its power from nuclear energy, Belgium aims at 45%, and France expects 70% of its electric power from reactors. The rate of new reactor construction has slowed in the U.S., but elsewhere is accelerating. The Soviet Union has 19 reactors and plans 18 more. Japan has 13 and plans another 15 by 1983.[2]

Despite international opposition that is no respecter of national boundaries, erection and use of light-water nuclear reactors appear likely to continue, unless a major disaster occurs somewhere fulfilling the grim warnings aimed at the industry by critics. Nuclear catastrophe might halt or slow the trend. Nothing short of catastrophe seems likely to do so. In March

1976, 38 countries other than the U.S. had 260 nuclear reactors ordered, under construction, or operating. Countries with an insatiable modern need for energy are still turning without hesitation to nuclear power as the best available option.

Trust it or not, nuclear fission has asserted a determined claim for recognition as a future energy. And it has already been widely recognized.

SCIENTISTS FOR, SCIENTISTS AGAINST

In a 1975 statement, 32 American scientists defended nuclear energy.

> We see the primary use of solid fuels, especially of uranium, as a source of electricity. Uranium power, the culmination of basic discoveries in physics, is an engineered reality generating electricity today. Nuclear power has its critics, but we believe they lack perspective as to the feasibility of nonnuclear power sources and the gravity of the fuel crisis. All energy release involves risks and nuclear power is certainly no exception . . . . As in any new technology there is a learning period.[3]

The defenders believed the safety record in the nuclear industry was a good one that deserved trust. They could see "no reasonable alternative to an increased use of nuclear power to satisfy our energy needs." These scientists, including Nobel laureate Hans Bethe (Physics, 1967), were prominent professionals.

But equally distinguished scientists lined up against them, including Nobel laureates George Wald (Physiology/Medicine, 1967) and James Watson (Physiology/Medicine, 1962). Nobel physicist Hannes Alfven (Physics, 1970) confronted nuclear proponents with this challenge:

> The technologists claim that if everything works according to their blueprints, atomic energy will be a safe and very attractive solution to the energy needs of the world. This may be correct. However, the real issue is whether their blueprints will work in the real world and not only in a "technological paradise."[4]

In 1975, on the thirtieth anniversary of the bombing of Hiroshima, over 2000 scientists signed the "Scientists' Declaration on Nuclear Power," containing this statement:

> The country must recognize that it now appears imprudent to move forward with a rapidly expanding nuclear power plant construction program. The risks of doing so are altogether too great. We, therefore, urge a drastic reduction in new nuclear power plant construction starts before major progress is achieved in the required research and in resolving present controversies about safety, waste disposal, and plutonium safeguards.

This statement was widely criticized, but it and other sober commentaries questioning the prudence of plunging ahead into the nuclear unknown did have effect. In 1974, 27 new reactors were ordered in the U.S., and only 5 the following year. The slower rate continued as the decade moved on toward the 1980s, but indications were that electric utilities would not drop nuclear plans entirely and that implementation of those plans might accelerate under the pressure of rising fuel costs. Utilities across the country argued that nuclear energy was the only practical option available, with oil too expensive, coal not available in the quantities required, and with solar energy unable in this century to contribute significantly.

Such conditions prevailing, nuclear energy seemed the logical and most economical alternative. "We'll need all the nuclear power we can possibly get by 1985. There's no other way in sight," wrote Keith Cunningham, President of United Nuclear Corporation. Cunningham noted in 1977 that oil and gas are too scarce and valuable for continued use as boiler fuels, which means coal and nuclear power must be rapidly increased. Countering this view was Barry Commoner who accused power companies of ignoring solar energy because their domination would be threatened if individual homes were energy self-sustaining. Commoner also doubted the claims that nuclear energy is as economical as claimed. According to Commoner, costs are swallowed up in the rates, and "the consumer bears the burden of technological foolishness."

The pro and con dispute on nuclear energy was an endless circle, like a carousel with bobbing horses fixed in place and the riders shouting ineffectually at one another.

Essentially what was taking place was a waiting game. Each side was waiting for proof, scientifically sound proof, that it was right. The chance was good that such proof would not be available on either side for many years. Mankind would have to go ahead with nuclear energy without assurance of safety. It would have to generate radioactive wastes in the expectation that some way would be found to dispose of them safely. Yet that expectation was inescapably subject to the realization that research to manage nuclear reactor wastes had been underway for years, and there was still no answer.

NUCLEAR FISSION AND THE QUESTION OF SAFETY

In 1977 the nuclear industry in the U.S. is 20 years old. The first commercial production of electricity from nuclear energy occurred in 1957. On December 20, 1951, at an experimental fast breeder reactor in Idaho, electricity was first generated and feasibility demonstrated.

The point is that nuclear energy has not had a century or even half a century to get the bugs out, prove its technology, and solve inevitable problems. Yet during those first 20 years, with over 60 reactors operating in the U.S. by the mid-seventies, few fatalities attributable to radiation have occurred at any of the U.S. reactors. This safety accomplishment is impressive, but of course it does not prove that nuclear reactors are "safe." It proves only that no accident to date of whatever seriousness has produced large-scale disaster, and it can be argued that a convincing record of past safety does lend credence to the probability of future safety.

A 1977 report from France noted that radioactive gas leakage from a nuclear fuel processing plant had been controlled before any damage was done. Mishaps in the U.S. have occurred, but without moving from crisis to widespread calamity. The record shows that despite inevitable mishaps and failures, nuclear facilities have been effective in quick response to prevent

NUCLEAR FISSION SCHEMATIC (Light-Water Reactor)

1. Finding Uranium: Leaves the mine, a bright yellow ore. Less than 1% can be used as nuclear fuel. This 1% is uranium-235 (U-235). Most of the ore is U-238. U.S. uranium reserves are estimated 1.8 to 3.7 million tons. Australia is another major source of uranium (20-25% of the world's reserves in non-communist countries).

2. Enriching Uranium: The fuel mixture used in a light-water reactor is 2-5% U-235 and the rest U-238. Uranium ore goes through an enrichment process to achieve this mixture. The ore is changed to uranium dioxide and pressed into fuel pellets, which are welded into 14-ft zirconium alloy fuel rods. Fifty rods make a bundle.

3. Nuclear Core: About 700 bundles are arranged together, forming the nuclear fuel core, cylindrically shaped, weighing about 150 tons.

4. Into the Reactor: The nuclear core is inserted into the reactor, which goes "critical" and fission begins. The core becomes extremely hot. The heat is carried by circulating water to drive a turbine and generate electricity.

5. Circulating Water: Every minute about 350,000 gallons of water flow through the core during operation. This keeps the reaction going and removes the heat. The water is kept from boiling by a pressure of 2000 psi. This coolant is essential to avoid meltdown (the nuclear disaster feared most).

6. Radioactive Wastes: During fission, uranium atoms are split and a number of radioactive substances form (strontium-90, cesium-137, iodine-131, krypton-85, plutonium-239). Such wastes eventually interfere with fission and must be removed. Disposing of wastes safely is a major, unsolved headache for the nuclear industry.

7. Spent-Fuel Pool: At 1- to 2-year intervals, the reactor is shut down to replace spent fuel bundles. Bundles near the center of the core become hottest and accumulate most radioactive waste. Fresh bundles are added to the core, while spent bundles, hot and radioactive, are submerged about six months in the spent-fuel pool.

8. Emergencies and the Spent-Fuel Pool: A spent-fuel pool normally holds 1-1/3 cores, but no more than 1/3 of a core should occupy the pool at one time. Thus, space is always available in case the active core in the reactor must be submerged and deactivated with emergency speed.

9. Water in 30 Seconds: In an emergency, if the core could not be submerged, meltdown would start. If water stopped flowing through the core, the reactor would "scram" as boron rods dropped into the core to stop fission. But nuclear wastes in the core would continue fission. Temperatures would pass 5000°F, and meltdown would start if water did not flood the core in 30 seconds.

10. Meltdown: In a meltdown, the bottom of the reactor would melt through, and the molten core would fall to the bottom of the containment structure. Superheated, radioactive steam would break through to the outside, and radioactive gases would spread through the area. The surrounding land would be saturated with radioactivity. Near cities, a reactor meltdown could cause 45,000+ deaths, cost $40+ billion in damages.

11. Recovery Plant: After cooling in the spent-fuel pool, bundles are supposed to go to a recovery plant, where wastes are separated for safe storage. Recovery plants so far have suffered frequent shutdown. Uranium is sent to enrichment facilities for use in new bundles.

12. Storage of Radioactive Wastes: A problem. Storage containers have been found to leak. An answer is still being sought. Burying plutonium deep in the ice of central Antarctica or firing it into space have been seriously considered. Plutonium-239 is a danger to man for several hundred thousand years. And it can be used to make nuclear weapons.

catastrophe. Luck may be involved, but it can't be all luck. The people involved must be good at their jobs.

In the 1970s nuclear critics have tirelessly stressed the hypothetical threat of a major accident at a nuclear site followed by a full-scale meltdown. But most criticism tends to focus on radioactive wastes, still mounting in quantity with no end in sight. Finding ways to manage safe long-term storage of wastes is a critical priority.

Biologist George Wald noted that 1 milligram of plutonium-239, when inhaled, will kill an individual in hours through massive fibrosis of the lungs, while 1 microgram would probably induce lung or bone cancer.[5] Yet plutonium still accumulates at nuclear sites around the world. Nevertheless, however imperfect the storage facilities and practices, plutonium and other radioactive wastes have been controlled. They have not been allowed to unleash a nightmarish nuclear plague on mankind. That too does not prove safety, but it suggests that responsible care has been reasonably exercised.

Obviously those arguing for a "go slow" policy on nuclear energy are not convinced by the bad things that have happened but by the terrible things that might. The staggering dimensions of the threat (thousands of deaths, billions in loss, decades for the land to cease being hazardous) in case of one serious reactor accident due to breakdown, earthquake, weather, sabotage, etc., bring insistence on what has never been demanded from any other energy system or human enterprise: absolute assurance of safety.

FAST BREEDERS

In the early 1970s, fast breeder reactors were considered a solution to the world's growing shortages of uranium. The breeder was expected to replace light-water reactors in whole or in part by the 1980s and serve as man's prime energy source for the 21st century and beyond.

The breeder sounded pretty good on paper. It received its name from the fact that it produces nuclear fuel in the process of consuming fuel and provides a self-sustaining energy source.

The breeder reactor, when/if it succeeded, would be a modern fulfillment of the ancient human quest for perpetual motion machines, totally self-renewing energies, and other fond dreams from fairyland.

The breeder would have several advantages over light-water reactors. It would not require the costly uranium enrichment and elaborate fuel cycle associated with light-water reactors. Plutonium bred in the breeder reactor would be relatively inexpensive and simple to convert into new fuel elements.

Work on breeder reactors went ahead in several countries. A $2.2 billion breeder reactor was planned for construction near the Clinch River at Knoxville, Tennessee. A fast breeder reactor, the Phenix, was built near Marcoule on the Rhône River in France. This prototype breeder reactor went critical in August 1973. Other experimental breeders were designed by the Soviet Union and Great Britain. In Canada, four CANDU breeder reactors powered the nuclear generating station near Toronto. The CANDU was an intermediate step between water and breeder reactors. It was hailed by some American scientists, including Dr. Bethe, as an impressive achievement.

So development went ahead on implementation of the breeder concept. In the U.S. nearly a quarter of a billion was invested in fabricating parts for the Clinch River Breeder Reactor at factories throughout the country. The French reported that the Superphenix, a second-generation breeder, would be operating by 1981. The Superphenix, benefitting from experience with the Phenix, would have improved breeding qualities, solve steam generator problems, and prove the economic feasibility of fast breeder technology.

But there was a major problem with fast breeder reactors, a familiar problem: Plutonium. With countries rapidly committing themselves to the breeder reactor approach, arguments grew that a dangerous gamble was involved. Even in minute quantities, plutonium was a known threat to the welfare and survival of terrestrial life. Breeder reactors by vastly multiplying the output of plutonium could at the same time vastly multiply the threat.

MAIN SAFETY PROBLEMS OF NUCLEAR POWER

Major Accident to the Reactor: Depending on the reactor location, the cost in lives and property could be gigantic, rivaling the worst natural calamities (tidal waves, earthquakes, volcanic eruptions, the Black Death of the 14th century).

Status: During the first 20 years of commercial nuclear enterprise, there has never been such an accident. "Near misses" stressed by opponents are claimed by supporters to prove that even when trouble develops, safety provisions in reactor systems take care of them. "We've been lucky up to now," say antagonists. "Risk can never be eliminated from life," supporters reply.

The safety record of reactors is good. To increase safety, the U.S. Nuclear Regulatory Commission is requiring more and better trained guards. Frequent unannounced inspections are to be made at light-water reactors, and a federal inspector is proposed as a permanent official at each plant. The siting of future nuclear plants will be based on safety criteria and will strive "to reduce the risks of a nuclear accident and the consequences should one occur."[1]

Plutonium: Plutonium-239, by-product of a light-water reactor, is extremely toxic, and serves as the "trigger" for a fission atomic bomb. A single reactor produces sufficient plutonium annually to manufacture several bombs. Approximately 7 kg of plutonium makes an atomic bomb comparable in power with the one that destroyed Hiroshima. This makes reactors vulnerable targets of terrorists, dissidents, and criminals. Keeping plutonium from being stolen, and keeping it from polluting the atmosphere are permanent obligations of the nuclear industry. They are difficult, perhaps impossible obligations to meet invariably and without exception.

Status: Security is continuous at a nuclear power plant, and it is steadily being tightened. Security failures inevitably have occurred and will again. No major mishap has yet happened. With hundreds of reactors in operation worldwide, plutonium protection is an international necessity that calls for international management.

Radioactive Waste Disposal: Disposal systems tried so far have proved inadequate.

Status: Some poisonous products of the nuclear chain reaction decay swiftly (such as iodine-131). Others, such as strontium-90 and cesium-137, capable of causing cancer, leukemia, etc., must be isolated from man. Plutonium, deadliest of all, requires isolation for at least 100,000 years. Burial of these wastes in tanks, as at Maxey Flats, Kentucky, has proved inadequate because of leakage. Such storage has been considered a stopgap solution until technology is perfected for safe disposition of wastes. Other methods of handling wastes include holding spent fuel rods in deep pools of water, preventing excessive radiation leakage. But such storage is limited. "The Consolidated Edison Company has enough space at its Indian Point plant in Buchanan, N.Y., only for the next eight years."[6] After that what?

The Nuclear Fuel Services plant, West Valley, N.Y., was designed to reprocess nuclear waste and make it safe for further use. In 1977 this plant had been declared a failure and was being dismantled at a cost that might reach $1 billion.[6]

In 1977, the future of the nuclear industry was still tightly entwined like a tangle of yarn with the crucial puzzle of waste disposal. If the problem could be satisfactorily and believably resolved, as many scientists including critics of nuclear power thought it could be, then the way would be wide open for nuclear energy to leap ahead. Until then, however, every slow advance would be difficult and made against massive opposition with dangerous wastes multiplying constantly.

117

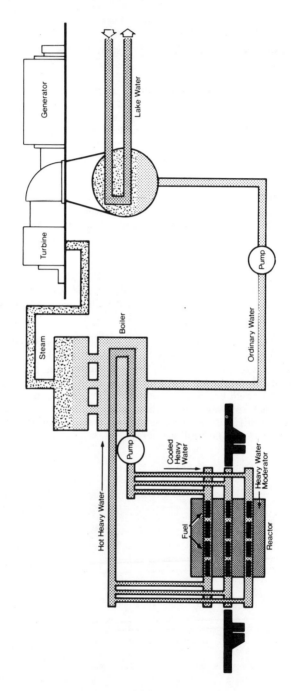

Figure 15. Schematic of the CANDU reactor. (From *Environmental Aspects of Nuclear Power* by Geoffrey G. Eichholz, Ann Arbor Science Publishers, Inc., 1976.)

118

The dangerous qualities of plutonium thus led to a U.S. Presidential recommendation in April 1977 not to continue with the Clinch River Breeder Reactor. The U.S. should refrain from stepping across the threshold into the perilous Age of Plutonium and would encourage other nations also to refrain. The U.S. approach was stated in the 1977 National Energy Plan:

> It is the President's policy to defer any U.S. commitment to advanced nuclear technologies that are based on the use of plutonium, while the United States seeks a better approach to the next generation of nuclear power than is provided by plutonium recycle and the plutonium breeder. At the same time, because there is no practicable alternative, the United States will need to use more light-water reactors to help meet its energy needs.[1]

Although the U.S. in 1977 might be delaying further progress on the Clinch River Breeder Reactor, vigorous research was going ahead on breeder technology with millions budgeted. The U.S. might hold back on immediate application instead of rushing ahead, but it wasn't dropping out of the game. U.S. atomic scientists would continue to seek ways of making the breeder work and work safely. Other countries, however, were not holding back, although the U.S. president and the Executive Branch of the U.S. government had declared against technologies such as the breeder reactor that might stimulate nuclear proliferation worldwide (spread the capacity for atomic bomb manufacture).

Some Americans thought it was an out-of-the-frying pan situation to cancel fast breeder reactors and then accelerate the proliferation of light-water reactors, themselves a source of radioactive wastes, though admittedly not breeders of plutonium with the spectacular aptitude predicted for the fast breeders. Others argued the dangers of turning away from fast breeder technology in the face of dwindling uranium supplies, the decline of traditional fuels, and the impossibility that reasonably priced electricity from solar energy could be available soon enough to meet the needs of the 1980s.

With a decade needed to plan and complete nuclear energy installations, every decision inevitably is a future energy decision. Decisions in 1977 have little consequence for the present, but great consequence for 1987. Those who call it a mistake to hesitate in connection with nuclear energy and to draw back from new, if dangerous technology want to know what energy will be available to users in the future. Good question.

The same argument is used by their opponents. By what right do we continue creating lethal wastes that will be handed to the future like a gift-wrapped time bomb? Why isn't it more sensible and compassionate to conserve energy, to use less, and to develop rational, safe energy systems we can pass to the future in benign forms rather than in deadly, fragile forms that require burial in the ice of Antarctica for 100,000 years to avoid killing all life on earth? Good question. Could both be right? And wrong? Will the future remain uncertain whether we play it bold or play it safe? All of these are persistent—and disturbing—questions.

AN ICY PROPOSITION

There is no lack of imaginative suggestions for getting rid of radioactive wastes. Some of the recommendations have the echoes of science fiction about them, but are usually meant quite seriously. Launching wastes into space and sending them off toward another, preferably distant galaxy is one. One of the suggestions that is at once among the most bizarre and at the same time the most practical appeared in 1972. A carefully thought out proposal was made to establish a "Permanent International High-Level Radioactive Waste Depository in Antarctica."

This may seem a dirty trick to play on a continent that has always minded its own business, never littered, and remained a good, quiet if cold and distant neighbor. Yet fair or not, burying radioactive wastes in the 2-mile-thick ice of the South Polar Plateau appears safer than burying it in Kentucky or elsewhere in the heavily populated Northern Hemisphere.

Operation of the Breeder Reactor

Fuel: Plutonium-239 or uranium-238 enriched with uranium-235.

Neutrons: Fast neutrons rather than slow neutrons are generated during fission (opposite is true in the light-water reactor).

Breeding Facility: Converts nonfissionable uranium-238 to fissionable plutonium-239. Manufactures fuel while producing usable heat in the reactor core. Effective energy from uranium is 60 times greater.

Coolant: Liquid sodium (instead of water). The fuel is inside thousands of thin metallic pins. The pins are in cans called subassemblies. The cans, hexagonal in shape, are tightly packed together on the reactor floor.

When control rods are removed, the reactor goes critical. Neutrons are released throughout the core and the surrounding blanket of uranium-238. Fission occurs, heat is produced. In the blanket, when a neutron is captured by uranium-238, plutonium is created, which can be used as a breeder reactor fuel.

Fast Neutrons Explain The Breeder Function: Fast fission delivers a surplus of neutrons. That surplus allows the fast breeder reactor to create more fuel than it consumes.

Secondary Sodium System: Liquid sodium circulates as a coolant through the breeder core. A secondary sodium system takes heat from the primary system and transfers it to water for steam production and subsequent turbogeneration of electricity. Turbines and generators are protected from radioactive sodium.

In the French Phenix, transferring heat from hot liquid sodium in the core to the water that produces steam has been a source of difficulties, with connections between the nuclear and the turbogenerator systems the main problem. According to an engineer, the problems are all in the plumbing.[7]

Efficiency: "It is expected that the breeder will be able to operate at a thermal efficiency of about 40% as compared with about 32% for current nuclear plants."[8]

In the Phenix, the French reported maintaining fission until 5% of the fuel had been used. If 10% of the fuel could undergo fission ("10% burnup"), costs for the fuel cycle could be cut in half. Greater fuel efficiency thus is a key goal of breeder reactor research.

Except for a few widely scattered scientific bases (*e.g.*, Americans at the South Pole, Russians at Vostok), the South Polar Plateau is lifeless. A wandering skua bird may occasionally fly over it. Explorers have reported seeing confused skua far inland. But neither birds nor any other living creatures are native to the forbidding climate of the Antarctic interior.

The South Polar Plateau has the space, the isolation, and a permanent freeze to ice over anything deposited there. C. B. Reed, an expert on nuclear waste management, writes:

> Admittedly, the logistics will be difficult. The project will also involve risks, for every time these high-level wastes are handled there is danger. They must be shipped by truck or rail to a seaport, loaded on shipboard and offloaded under difficult conditions. They must then be transported a thousand miles to the bleakest land in the world before the ice can engulf them. But these risks, in my judgment, are warranted if we can build there a permanent *disposal* system for the wastes we have and the wastes that will be forced upon us. Our alternatives are few and all are hazardous. In the long run, the most hazardous is the current plan to store high-level wastes in our own back yards for hundreds of years.[9]

Getting wastes to the Antarctic, a singularly austere and inhospitable desert of ice throughout the interior, would be difficult but manageable. The deadly materials would be stored in cylindrical or spherical steel cans, each containing 7.5 ft^3 of radioactive wastes. They would be flown inland during the brief Antarctic summer when the sun provides daylight and dropped on the ice cap at intervals of 10 km^2. The containers would gradually sink through the ice and be covered. According to Reed, "The evidence given warrants the conclusions that the cans will come to rest on the rock and the ice cap will safely cover the radioactive wastes for at least 100,000 years. If the incredibly difficult sea, land and air conditions were overcome and all the proposals carried out as planned, then 62,000 cans containing about 465,000 ft^3 of wastes could be emplaced in the proposed depository."[9]

Scientists from many nations have proved man can function in the Antarctic and that a project of this magnitude is not impossible. Richard E. Byrd began preparing the way for burial of man's lethal garbage when he and his companions flew a tri-motor aircraft over the South Pole in 1929. That action in effect claimed the Antarctic as a gigantic continental laboratory for science. Depositing scientific wastes there was never what Byrd had in mind when he more than any other man colonized the Antarctic with volunteer explorers and scientists for the sake of terrestrial knowledge and research. But if Byrd could have been convinced of the need and the logic, he would have helped plan the logistics, which will be formidable if this astonishing scheme is ever carried out.

That may not happen. The plan has a fictional aura. But we are long past being surprised when fiction and reality meet at the crossroads where technology catches up with make-believe. There is no fact more real than the necessity of putting radioactive wastes where they can do no harm.

URANIUM ENRICHMENT

In May 1977 President Carter approved shipment of approximately 827 pounds of enriched uranium to Belgium, Canada, Japan, Netherlands, and West Germany for research applications. This modified an American reluctance to distribute enriched uranium extensively (another safeguard against the danger of proliferation). Making enriched uranium available was a compromise offered to persuade other countries to cooperate in the American effort to curb proliferation.

As a substitute for the uncertain Clinch River Breeder Reactor, an even more expensive uranium enrichment plant was proposed for Oak Ridge, Tennessee. The plant, instead of using the traditional gaseous diffusion technology, would apply the new, more energy-efficient gaseous centrifuge technology. Reportedly the new technology would accomplish uranium enrichment using less than 10% of the electricity needed for gaseous diffusion, and deliver enriched uranium more economically.

If the U.S. lead in emphasizing breeder reactor research rather than hasty construction of breeder reactor installations takes effect in other countries as well, the expectation is that light-water reactors will spread rapidly. When this happens, greater supplies of enriched uranium will be essential. As an encouragement to other countries in resisting dangerous reliance on plutonium-intensive technologies at this stage in man's effort to climb the nuclear mountain, the U.S. may try to help meet the need for enriched uranium more actively and generously than before.

AHEAD FOR FISSION?

When did the atomic age begin? Did it start December 2, 1942, on a squash court at the University of Chicago when Italian physicist Enrico Fermi led a research team in building an atomic pile and achieving the first nuclear reaction? At Fermi's direction, control rods were slowly withdrawn from the atomic pile. The control rods were made from cadmium to absorb neutrons and prevent a chain reaction. Overhead were three men, a so-called "suicide squad," with liquid cadmium to flood the experiment if it began running away, if it showed signs of blowing up Chicago, or the earth, or the universe.

Nothing was blown up, of course. The experiment went off without a hitch, and the recording apparatus showed a chain reaction taking place in a controlled atomic pile, just as predicted. American scientist Arthur Compton called fellow scientist James Conant to report: "The Italian Navigator has reached the New World." Conant wanted to know how he found the natives. The answer: "Very friendly."[10]

In the 1970s, how the Italian Navigator found the New World was still being debated and assessed. Was atomic energy creator or destroyer? Obviously it could destroy with agonizing flair and had. For that reason, was it more accurate to pick another day for the start of the atomic age: A day in August 1945 when the Enola Gay, American B-29, dropped an atomic bomb on Hiroshima, Japan, and helped bring a savage war to an end by destroying the city with a single force equal to 20,000 tons of dynamite.

Or did the atomic age have its start on the second day of August, 1939 when Albert Einstein sent a letter to President Franklin Roosevelt, a letter which caused political wheels to turn and technological funds to appear.

124

> . . . In the course of the last four months it has been made probable—through the work of Joliot in France as well as Fermi and Szilard in America—that it may become possible to set up a nuclear chain reaction in a large mass of uranium, by which vast amounts of power and large quantities of new radium-like elements would be generated. Now it appears almost certain that this could be achieved in the immediate future.[10]

Einstein's knack for accurate prediction was no less in 1939 than earlier in the twentieth century when he was predicting the outcome of astronomical measurements based on the Theory of Relativity. Vast amounts of power were and are generated. The atomic age, like it or not, has arrived. The result is a number of troublesome energy options that we need. We are afraid to adopt these options. We are afraid not to adopt them. And these opposing fears represent where we are in the 1970s with some going one way, some another.

Despite fears, atomic developments are inching ahead. Though slowed by internal troubles and external opposition, nuclear development continues to survive and advance, but haltingly like a whale hampered by the harpoons of a thousand angry Captain Ahabs with busy crews. The fact that in 20 active years of nuclear energy development there have been no major catastrophes and few deaths caused by radiation is a remarkable record that stands as an indication of stringent safety measures and exceptional caution. In 1974, L. Manning Muntzing, director of regulation for the Atomic Energy Commission, said that nuclear facilities would have to be built and operated "with a discipline that has not normally been required in American industry." The safety record of the industry indicates that this relentless standard has largely been met.

GUARDING THE FUTURE

The fact that various reactors have experienced considerable down time is a further proof of inflexible safety standards. The record indicates that safety, not economics, is the prime consideration. Nevertheless, the nuclear industry has been under attack not for what it has done, but for what it might inadvertently do. In other industries, such hypothetical arguments would be shrugged aside with a "go away, don't bother me" gesture. Yet because of the size of the ultimate threat, such efforts to dismiss would exacerbate opposition in the nuclear industry.

Recognizing the potential dimensions of an atomic reactor disaster, we do not seem ill-advised to require special caution in connection with nuclear energy. There have been no disasters ("yet" the critics might say), but even a near miss inspires renewed commitment to uncompromising vigilance. John G. Fuller's book, *We Almost Lost Detroit,* reports on the occurrence of critically high radiation levels at a nuclear reactor near Detroit that qualifies in the "near miss" category. The danger of an explosion existed yet, again, it did not happen. Other nuclear mishaps have been reported in the news media.

So it is no surprise that warnings, regulations, demonstrators, and operational complications have kept atomic power momentum to a sluggish pace and frustrated efforts to accelerate. Caution is obligatory. We would be foolish to consider this regrettable or to invite a modification of caution. When Fermi and his associates withdrew their cadmium rods and ushered the human race across the threshold of atomic energy, they did so with utter care, absolute watchfulness, and complete respect for the tremendous powers involved. The liquid cadmium overhead was only one mechanism of diligent tribute. Those scientists knew, and so should men today and in the future, that with such colossal and primeval energies, to move without total caution is total folly.

With the dilemma of waste still unsolved, it does not seem responsible to increase the quantity of waste without establishing some specific way to handle it safely. In 1974 the U.S. nuclear industry produced 8000 pounds of plutonium-239. If nuclear expansion meets its growth predictions, about 600,000 pounds of plutonium will be produced annually. One ounce of plutonium, we are told, could form 10 trillion plutonium-dioxide particles, and if released could set off an international lung cancer epidemic.

Figure 16. Floating nuclear power plants such as this one may be developed at offshore sites in the future to avoid population centers. Advantage: Plenty of cooling water. Drawback: Vulnerability to storms and ship collisions. (From *Environmental Aspects of Nuclear Power* by Geoffrey G. Eichholz, Ann Arbor Science Publishers, Inc., 1976).

It hasn't happened. Safety precautions keep it unlikely to happen. But the sheer dimensions of the threat make it impossible to ignore. Could a madman poison mankind with a pinch of plutonium? Could political enemies sabotage a country through its nuclear facilities? What can be done to limit our vulnerability in the atomic age while benefitting from the tremendous energies of the usually peaceful atom?

Answering the technical, moral and political questions bound up with nuclear energy will be continuing concerns in the future. Increased understanding and control to eliminate dangers will be important goals. Caution may and no doubt should guide future use of nuclear fission, but there is no likelihood that it will not be used. Man's ability to extract energy from the atom has been demonstrated. It can be extracted economically and with reasonable regularity. Some countries, harassed by energy needs, are certain to continue driving the nuclear road, even if there are known hazards ahead, dangerous turns, and precipitous plunges to disaster if something goes wrong or a mistake is made.

The nuclear industry is troubled in the 1970s, sometimes beleaguered, exposed to challenge, subject to question. But it is still with us. It will still be with us in the 1980s. Energy from the atom is a marketable commodity. It will be released. It will be sold, unless something goes terribly and very noticeably and quite suddenly wrong. We can decide on a date when the atomic age began, but we don't know when it will end, or if it will end, or how it will end. It should work out all right if we stay cautious, advance with care, and get lucky about that cantankerous, stubborn, energetic, sinister plutonium.

NUCLEAR FISSION - REFERENCES

1. The National Energy Plan, Executive Office of the President, Energy Policy and Planning, April 29, 1977, pp. 69-73.
2. Jasen, Georgette. "Global Report," *The Wall Street Journal,* April 18, 1977, p. 6.
3. *Bulletin of the Atomic Scientists,* "No Alternative to Nuclear Power," 31:4-5, March 1975.

4. Gordon, Howard and Roy Meador. *Perspectives on the Energy Crisis,* Vol. 1, (Ann Arbor, Michigan: Ann Arbor Science Publishers, 1977), p. 4.

5. Wald, George, "There Isn't Much Time," *The Progressive,* Vol. 39, No. 12, December 1975, p. 22.

6. Severo, Richard. "The Nuclear Drive is On," *The New York Times,* June 26, 1977, p. 8E.

7. Metz, William D. "European Breeders (II): The Nuclear Parts Are Not the Problem," *Science,* Vol. 191, No. 4225, January 30, 1976, p. 368.

8. Eaton, William W. *Energy Technology,* ERDA, Office of Public Affairs, 1975, pp. 22-24.

9. Reed, Charles B. *Fuels, Minerals, and Human Survival,* (Ann Arbor, Michigan: Ann Arbor Science Publishers, Inc., 1975), pp. 45-46.

10. Fermi, Laura. *Atoms in the Family* (Chicago, Illinois: The University of Chicago Press, 1954), pp. 190-199.

CHAPTER 8

THE SOLAR FAMILY

Down in the valley, valley so low,
Hang your head over, hear the wind blow.
Hear the wind blow, dear, hear the wind blow;
Hang your head over, hear the wind blow.

American Folk Song

When we speak of solar energy, we commonly refer to sunlight and the various ways of using it for cooling, heating, electricity, etc. But solar energy properly should be considered a family name, with a number of important, versatile and unique characters in the family as well as sunlight. Solar energy might as accurately be thought of as a consortium of energies, united under the supreme corporate banner of the sun, or a team of energies with each member standing by and ready to go into the game when needed.

The family analogy fits best because each different mode of energy has the same parent, Big Daddy Sun. It is a cooperative family, of course, a family of symbiotes with little or no sibling rivalry and a readiness to assist and supplement one another. Sunlight gives up the struggle at sundown and is nothing but grateful if wind promptly takes over to carry on the family tradition. In fact, an increasingly usual practice by those who employ the solar family to supply energy is to divide the job up among different solars. The results of such practice are encouragingly effective.

To understand the full range of future energy prospects, it pays to consider each member of the solar family, because any one of them or all of them may do the family proud and achieve energy distinction.

WIND

An Elderly Solar Energy Returning for Another Blow

New technology and needs bring shifts in emphasis and expectations. To the surprise of some energy watchers, wind power has caught up with and perhaps is surging ahead of ocean-thermal plants, bioconversion, and photovoltaic power systems as a ready-for-use potential energy alternate.

Some might argue that the plow best symbolizes the settling and civilizing of a primitive land, as in the case of North America during the nineteenth century. But perhaps a better case could be made for a more elaborate and less inherently immobile apparatus: The windmill. With its readiness to do work whenever a stray breeze appears, it also symbolizes energy in use. On the western plains of America the windmill meant water for life. The large-vaned windmills of Holland still pump sea water from flooded lowlands. The name came from the milling of flour with the energy of the wind, a practice still followed in some parts of the world.

Wind is another of the permanent and inexhaustible energies because it too comes from solar energy, and it is now seen as a potentially valuable energy source for broader duties than milling and pumping.

Using the energy of wind to help man do his work is nothing new. Wind has been used in practical ways for thousands of years. Sailing ships, driven by wind, gave man his ability to circle the earth and in time to conquer this planet. The story of man under sail is one of the most exciting and important historical epics we have, and it was repeated innumerable times wherever there were oceans to cross and prevailing winds for propulsion. Invention of the sailing vessel gave man mastery of the oceans through use of a remarkable solar energy that would move him almost anywhere he had the nerve and the desire to go.[1] When Europeans, aboard sailing vessels, reached the Pacific islands in the sixteenth century, they found that most of the islands were inhabited by Polynesians. How had these people, obviously of a common racial ancestry, crossed thousands of ocean miles to settle widely separated islands? The answer wasn't simple. It involved awesome maritime knowledge, human tenacity, great sea canoes, plus a brilliant understanding and use of wind energy. And the Polynesian voyages took place hundreds of years before Columbus discovered the Americas, by putting wind to work.

Windmills aren't as old as sailing vessels, but they too have a long history. Engineers in the eleventh and twelfth centuries made progress in the development of water mills and tidal mills, but their main work went into windmills. Windmills became so numerous and profitable, Pope Celestine III (1191-98) took the inevitable step that always follows success. The Pope put a tax on them.[2]

Hundreds of windmills were on the job in northern Europe during the Middle Ages. But the windmill was old even then. There are records of windmills in Persia 1,500 years ago.

Thousands of windmills—an estimated 175,000—were erected in U.S. rural areas to pump water, mill flour, and other vital energy chores. In earlier decades of this century, windmills were even used to charge electric batteries. Rural electrification and other convenient energies brought the gradual abandonment of windmills with inexpensive electricity available at the flip of a switch, but now a windmill renaissance may have begun.

At New Mexico State University College of Agriculture a course in windmill repair and upkeep has been introduced on request. In the last century, a professional "windmiller" was in demand. Then the world seemed to pass the windmiller by, but it may be catching up with him again. As the price of conventional energies moves up, the wisdom grows of not allowing a stray breeze to cross the prairie without getting some work out of it.

The Good Megawatt Provider

The use of wind power to generate electricity has been studied by the U.S. Federal Power Commission and reported to be economical for power plant capacities between 5 and 10 megawatts. Wind generators are used with storage batteries to provide electric power for houses. Ambitious wind turbine projects have been successfully operated in Russia, Denmark, Germany, France, the U.S. and elsewhere. Combinations of wind generators and solar collectors are recommended as a collaborative power venture.

> Winds are often most vigorous in cloudy weather when photovoltaic array outputs would be minimal and solar-thermal power systems would not work at all. On the other hand, on clear, calm days plants using sunlight would function best and wind plants not at all. If a large number of generators using sunlight and wind are dispersed over a large geographical area but connected to the same power grid, storage requirements would be minimal.[3]

Since the energy crisis of 1974, efforts to perfect wind generators as alternate energy sources have intensified. Experimental 100-kilowatt wind generators have been placed in operation with much larger generators down the road as experimental lessons are mastered and wind utilization technology improved.

Horizontal-axis wind machines are the common design approach, but radical new designs, including vertical axis designs, are being researched as well. Wind power is now recognized as a ready-for-the-taking energy source, available conveniently in many areas of the U.S. and demanding relatively inexpensive and uncomplicated equipment. As with any technology, reducing costs to make wind energy competitive with other energies is an immediate goal. Because wind is not continuous, mechanisms for energy storage will be essential. Energy storage is a challenge in connection with energy from many sources; and it may be that when the storage problem is solved, the solution will fit wind, as well as other sources. Meanwhile, as Williams suggests, using wind in conjunction with solar-thermal systems connected to a power grid should serve well.

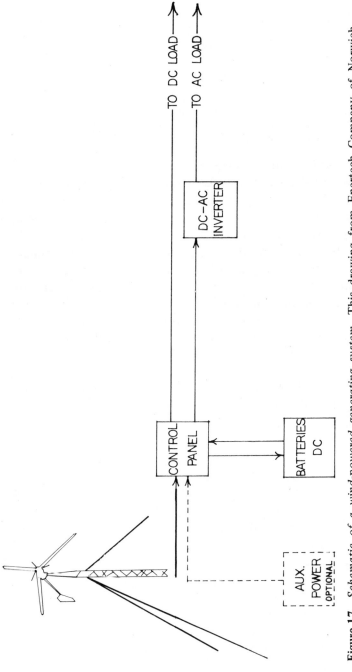

Figure 17. *Schematic of a wind-powered generating system.* This drawing from Enertech Company of Norwich, Vermont, shows how a wind turbine delivers power to a battery system which stores energy during high-wind periods for use in low-wind periods.

135

Efficiency figures are not discouraging but, for large amounts of power, wind turbine blades would have to be of a size sufficient to give Don Quixote delusions of grandeur.

> The efficiency actually achieved by a well-designed turbine is about 45%, and an overall system efficiency of 30-40% for generation of electricity can be obtained. Thus the blades of a turbine intended to provide 1 MW of power from a wind of 18 mph would have a diameter of about 180 ft.[4]

The environmental effects of installations for energy from wind are not expected to be severe, but there will undoubtedly be such effects, some of them negative. Noise could be a problem. Surface wind patterns might be altered by numerous wind turbines with large blades in busy operation. This might, the same as large arrays of solar collectors, change the microclimate in a local area, making it difficult or impossible for certain insects, animals and plants to continue surviving there.[5] Aesthetic objection to wind machines in the neighborhood probably will be encountered, as well as objections to potential radio and television interference. These are minor problems that should add up to little as environmental indictments of wind energy.

Areas of most effective energy production from wind in the U.S. are coastal regions where sea and land temperature differentials create extensive wind activity, and the Great Plains areas of the midwest and southwest.

Estimates by the World Meteorological Organization indicate a world total of 20 million megawatts is available through utilization of wind power. Approximately 350,000-megawatt generating capacity is estimated for the U.S.[6] Less conservative estimates of potential wind power range as high as 80 trillion kilowatts. The wind clearly delivers considerably more solar energy than is needed simply for stirring autumn leaves, teasing skirts, and clearing the air.

Winds averaging 12 mph are widespread in the U.S. Such velocities are believed adequate for economical production of electricity. A wind power enthusiast, William Heronemus, Civil Engineering Professor at the University of Massachusetts, has said, "If you admit that we must have a certain level of energy to keep our civilization going— and I myself don't see how we could reduce our current average usage by more than 20%—then we must have additional power plants, or, hopefully, new types of power plants that will remove the pollution caused by those now in operation. If one accepts that, and then realizes that the winds in certain wind-swept regions could supplant all of the existing fossil-fueled or uranium-burning plants, then shouldn't one be far more willing to look at windmills once in a while?"[7]

Figure 18. Wind generators of many types are available, and research continues to achieve greater power efficiencies. This modern wind generator is an example of those in current use wherever winds of sufficient velocity are encountered.

Bent Sørensen, a physicist at the Niels Bohr Institute, University of Copenhagen, in an assessment of energy prospects for Denmark, calculated that solar energy and wind energy combined could supply the country's energy needs by 2050. Sørensen advocated commencement of a shift to these alternate energies now, gradually and methodically.[8]

A reporter on wind power developments, Gary Soucie, wrote that when an abyss is reached, the safest direction is a step backward. Soucie holds that if wind were a dramatic, new, untried idea, "utilities might be falling over themselves to try it out. Instead they are knocking themselves out, and maybe the rest of us along with them, trying to make light-water reactors work . . . . Nothing alarms a technofreak so much as to suggest that he look backward."[7]

Looking backward has become one way of looking forward for serious energy watchers. That special and ubiquitous solar energy, wind, is receiving much more than casual looks. It is accepted as an energy idea whose time has come, again.

OCEAN-THERMAL ENERGY

Claiming Heat from the Ocean Surface

If you have gone swimming in water exposed to sunlight through the afternoon, you've made a valuable scientific discovery. You've learned that the upper water is warmer than the water below. Augment old swimming hole heat figures with those from the surface waters of oceans and other large water bodies and it is obvious the amount of heat available in ocean-thermal power is tremendous. Today thermal power plants have been designed for harvesting solar energy from the oceans. The process might be called "fishing for power."

The practicality of constructing power plants based on ocean-thermal power is taken quite seriously in a world now beginning to look at its energy options with the blinders and the conventional complacencies removed. Figure 19 gives one version of an underwater power plant complete with fish, boat, and a shining sun.[9]

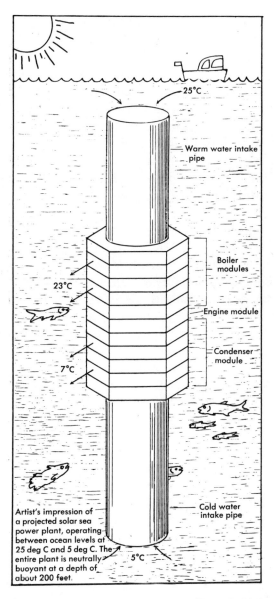

Figure 19. Modular ocean-thermal power plant. (From J. Richard Williams, *Solar Energy: Technology and Applications, Revised Edition,* Ann Arbor Science Publishers, Inc., Ann Arbor, Michigan, 1977.)

Ocean surface heat is another permanent source of solar energy, and in the large economy size at that. The heat in the Gulf Stream, for instance, has been estimated at 200 times the total power needs of the U.S.[6] Constant replenishment by the sun assures that this energy is never exhausted. Direct sunlight steadily heats ocean surfaces, while ice and snow melt in Arctic and Antarctic areas, creating currents of cold water moving far below the surface in the direction of the Equator. It has been estimated, according to Williams, that "the tropical oceans in the year 2000 could supply the whole world with energy at a per capita rate of consumption equal to the U.S. per capita rate in 1970 and suffer only a one-degree C drop in temperature."[3]

Taking advantage of ocean-thermal layers for energy was proposed by Jacques d'Arsonval as long ago as 1881. Modern energy needs have inspired practical new schemes for using this plentiful energy. One system outlined for the International Solar Energy Society and printed in the *Congressional Record* was developed by James Anderson and his son.

The Anderson method requires a floating power plant with a circulation system that works as follows. For a plant of 100,000-kilowatt capacity, a 35- to 40-foot cold water pipe brings cold water from depths of about 2000 feet. Warm surface water is pumped through boilers where it evaporates liquid propane. Propane vapor goes from the boilers to the turbine, where it expands and delivers energy to drive the turbine and generate electricity. Propane returns to the condenser, is condensed and liquefied by cold water, then repeats the cycle. And it is a cycle that can go on indefinitely. Ocean-thermal energy is one of the solar renewables.

A comparable plan proposed by Dr. William Avery, Johns Hopkins University Applied Physics Laboratory, uses liquid ammonia in a similar circulation system. According to Avery, six-sevenths of the total sun energy reaching earth is absorbed by the ocean's upper layer, making the ocean a particularly valuable source of solar energy in a concentrated, manageable form.[10] The U.S. National Science Foundation has financed ocean-thermal studies, concentrating mainly on propane or ammonia in a warm water-vaporization-cold water system of the sort described. A program to acquire this energy has been recommended, with plants moored near Miami and electricity transmitted ashore by undersea cable.[11]

Energy storage is not a problem with ocean-thermal conversion because the energy is continuously available. That means potentially economical operation of an ocean-thermal plant, but this fact must be balanced by the high cost of construction and maintenance of facilities at sea. Estimated capital costs for such plants have been placed at $200-$300/kW, which is competitive with fossil fuels and nuclear energy. With rising prices for the latter, ocean-thermal energy might come to offer a price advantage.[12]

Environmental and Ownership Questions

Ocean-thermal installations are certain to have an environmental impact wherever they are located. The sea life in the vicinity can't help but be affected when an "energy boom town" suddenly takes over the neighborhood. Whether rents will go up or down is hard to guess. The environmental results may be largely good ones, as cold undersea waters bring depth nutrients to the surface. Fish may thrive. Sea serpents may flourish. Neptune and his entire court may establish permanent residence nearby. The Flying Dutchman may anchor in the vicinity. On the other hand, the only measurable result may be the electricity generated from the warm waters and the cold. Byron in *Childe Harold* wrote:

> Roll on, thou deep and dark blue Ocean-roll!
> Ten thousand fleets sweep over thee in vain;
> Man marks earth with ruin-his control
> Stops with the shore.

The poet's assumption about the sea is sadly dated now. Man has asserted his control or at least his talent for attacking the ocean with everything from raw sewage to air pollution to mammoth oil spills. Offshore rigs now populate many ocean areas to take petroleum from under the sea bottom. Factory ships with giant dredges are now able to scoop up the sea floor in search of minerals, pirate's gold, or perhaps refuse from the Mayflower. The ocean floor is supposed to have sufficient copper for 6,000 years, nickel for 150,000 years, aluminum for 20,000 years, compared with land reserves for each of no more than a century. With such wealth out there waiting, man's control is no longer going to stop with the shore. You can bet on it.

But who does control the ocean and the ocean floor? The United Nations has held conferences on the subject. Others will be held, as more men go down to the sea in more ships equipped to find the wealth and scoop it up.

> More than half of the world lies two miles deep beneath the sea. For a generation, in increasing numbers, men in small ships have prowled the oceans, profiling, photographing, and sampling the floor of the sea, drawing the shape of a landscape no one had ever seen, finding new mysteries.[13]

The little ships will become big ships, and the mysteries will be explained. The aluminum will be harvested. How much such activities will upset the sensitive ecosystems of the marine world no one knows for certain. Electricity certainly will be among the products gathered from the sea. It may turn out that ocean-thermal plants do no harm at all while stirring the ancient waters. Taking solar energy from the ocean, which it has so plentifully, shouldn't interfere with another of Byron's poetic portraits of the sea.

> Dark-heaving—boundless, endless, and sublime,
> The image of eternity.

It is an image to cherish. From the viewpoint of future energies, the notion of an eternal ocean offers us the consoling prospect of eternal energy as well. There it is, another in the solar family, everywhere at sea, waiting for us to get around to it.

LIVING ENERGIES FROM ENERGY FARMS

Solar energy offers a means of replacing fossil fuels in an intriguing manner. Clean, renewable fuels are provided, and we do not have to wait 40 million years for delivery.

142

What's the secret? Take a look at your garden, or if you've been too busy recently, at your weeds. The secret is told in one word: Photosynthesis.

The method in essence would be a calculated, systematized variation on the familiar backwoods business of going into the forest with a hatchet and bringing back wood for the fireplace. Plants are past masters at the technique of using solar energy to manufacture carbohydrates in the photosynthesis process. Why not cultivate specially selected plants on a large scale and burn them for electric power?

An "energy plantation" would supply the plants on a continuing basis. The economic feasibility of an energy plantation has been argued with figures indicating such a plant would be no more expensive to construct and run than a fossil fuel steam-electric plant.[14] The NSF/NASA Solar Energy Panel considered this method as a serious possibility and concluded that the power costs at a plant using trees might fall between $1.50 to $2/MBtu.[15] The same cost estimates were given by the panel for the suggested technique of producing methane from algae grown in sewage ponds.

Employing solar energy to grow plants, and then burning plants to meet power needs can hardly be labeled a new energy source. The same method was employed by whichever of our ancestors regularized and domesticated the use of fire. Yet even if it could be called progress in reverse, the method is recognized as another practical source of future energy.

An oil well or a coal mine could be called an energy farm without belaboring the idea too violently. In each, the plants were simply left there several million years ago instead of last year.

Fast growing trees and other plants can be burned to produce electric power, or they can be processed to produce fuels such as alcohol or methane. This objective is stated succinctly in the 1977 National Energy Plan: "Agricultural and forestry residues already are used as fuel, and that use can be increased by improved collection methods and by energy farms, in which crops are grown specifically for use as energy."[16] Philip Abelson explains what is involved in more detail:

143

There has been much talk of solar energy, but most thought has been devoted to physical means of collecting sunlight. An obvious resource has had only nominal attention: energy materials, and chemicals from plants and trees .... Ultimately, trees may be the preferred source of energy. However, effective use of them for other than direct burning involves complex technology. Cellulose and lignin must be separated and then processed further or the wood must be converted to carbon monoxide and hydrogen and from thence to methanol or methane.[17]

Plants suggested for growth as "biomass" include sunflowers, rubber plants, and eucalyptus trees which grow fast and thus deliver maximum energy. Nevertheless, the energy efficiency of agriculture is poor at the present time. Only a minute percentage of the solar energy consumed is delivered in the form of biomass. Photosynthesis efficiencies can be improved by developing crops that achieve maximum solar energy conversion while losing a minimum quantity of energy through respiration.

In a hungry world, a question that persistently appears in connection with proposed energy plantations is whether the plants are needed more for energy or for food. The answer may continue to be juggled back and forth between the two demands for a long time to come, since growing populations will continue to be hungry for both food and energy.

Marine Energy Farm

Kelp farms at sea are being studied as a way to burn our plants and eat them too. Use of the ocean for energy farms would leave land areas free to grow crops for human and livestock consumption. Seaweed at sea and strawberries ashore sounds like a worthwhile arrangement. Kelp or seaweed is the earth's fastest growing plant, and harvesting it at sea in ocean farms for energy conversion is believed a realistic future energy possibility.[18]

One proposal entails harvesting kelp off the California coast, bringing it into port 300 tons at a time, moving it to a processing plant inland, and there converting the kelp to methane using a biological digestion process. Food and fertilizers would be by-products.

Starting with an experimental farm a mile wide, eventual kelp farms up to 100 miles wide are projected. The U.S. Navy and the National Science Foundation have supplied funds to study this potential energy.

Environmentally, marine farms are considered quite benign in their ocean environment, providing unexpected oases of greenery amid the rolling waves. Ashore, transporting and processing the kelp would write a different story with pollution and other problems to complicate the plot. Energy plantations on land require large areas, considerable water, and perhaps fertilizers that are often in short supply. Such plantations might also commandeer land that otherwise could productively be used for traditional agricultural purposes. Even with these hypothetical criticisms, future efforts to develop all possible energy sources may find bioconversion on land practical and justifiable. There are massive land areas available that are currently being put to little use. Some of these areas, with an investment of agricultural development energy, might in time deliver substantial payoffs in biomass energy.

Fueling Automobiles with the Brazilian Cassava

Brazilian scientists, seeking better ways to use solar energy, concentrated on the possibilities of plant life in their equatorial country. Their attention soon focused on the native Brazilian cassava plant, best known outside Brazil as the source of tapioca flour. In Brazil, the cassava has many uses. Grown by Brazilian Indians as a staple food, it is also fermented for alcohol, and processed to manufacture flour, cotton sizing, or laundry starch. Add one more product to the versatile cassava list: energy. The plant will be grown on a large scale, and fermented to produce ethyl alcohol. Among other energy applications, alcohol could be used as an automobile fuel with several nice features—absence of sulfur, low combustion temperature, and almost pollution-free exhaust gases. How much energy would be available from the cassava harvested for bioconversion? According to reported calculations, Brazil's energy needs could be met by using the cassava from 1% of the land area.[19]

145

The use of alcohol as an automobile fuel in place of or mixed with gasoline poses no serious difficulties to automobile industries. New machines could be manufactured at a cost of about $20 extra to fix the carburetor for efficient operation on alcohol. Currently, this form of fuel is twice as expensive as gasoline and gives half the mileage. With dwindling oil supplies and more economical production of alcohol, possibly from bioconversion, this disparity should eventually disappear. At that time automobile drivers probably will also find that their reluctance to turn their cars into alcoholics has disappeared. In fact, some marketing genius will proudly adorn one of his models with an inspired name, "The Alcarholic," and capture the imagination of the public.

BIOGAS

Disposing of waste is as old as the human race. The first human beings, the same as their descendants (us), must have had refuse, something they couldn't use and wanted to get rid of. Our primeval ancestors probably solved their problem of refuse simply by walking away from it. The world was large and uncrowded. It was easy to move on. When fixed residence living became the thing done, this process had to be reversed. The refuse had to be burned, carried away, buried, or left in your neighbor's yard secretly at midnight. Burning has perhaps been the commonest means historically for disposing of refuse in order to make room for more.

But what is waste? According to E. R. Pariser, M.I.T. Senior Research Scientist in Nutrition and Food Science, waste is "matter out of place." Pariser and other scientists want the energy now being lost in the 125 million tons of urban solid wastes thrown away annually in the U.S. to be salvaged. In addition to biogas, which can supply energy, thousands of tons of other reusable materials are present in the nation's refuse. Recovering precious resources from energy to scarce metals is a relatively new manner of looking at waste matter, but in an era of growing shortages and concern about energy and resources the new attitude may be permanent.

146

The challenge of resource recovery is as mammoth as the mountains of municipal refuse delivered daily to the designated disposal site. Different areas are trying to meet the challenge in different ways.

In 1973 Baltimore, Maryland, began constructing a Pyrolysis Resource Recovery Center. The plant will treat waste using the pyrolysis method—wastes are baked to convert them to biogas and carbon. The Baltimore plant uses the Landgard system developed by the Monsanto Company, in which gases are burned in an oxygen-deficient atmosphere. This is the last stage in the conversion process which begins with shredding wastes, followed by pyrolysis in a rotary kiln. The plant capacity is 1000 tons of waste per day.[20]

At the Hackensack Meadowlands in New Jersey, 50,000 tons of New York City refuse, normally dumped as landfill, will be shredded and converted to eco-fuel for use in producing electricity.

Even the U.S. Army has cast a thoughtful eye on waste for the sake of energy. Research studies by army scientists found that glucose and then ethyl alcohol could be produced in quantity from discarded newspapers (2000 pounds of glucose from 4000 pounds of newspapers). Watch it with that waste basket, soldier! It contains energy.

At a pilot plant in Kansas, Gulf Oil demonstrated that municipal waste can be converted into ethyl alcohol by the glucose process (500 pounds of glucose from a ton of waste materials).

Estimates concerning the amount of energy available from municipal refuse vary. The technology in this area is still being developed. Theories rather than proofs abound. But the Environmental Protection Agency, which has taken the lead in the U.S. in promoting the development of technology for energy and resource recovery from wastes, has estimated that daily refuse from 70% of the U.S. population contains energy equivalent to 500,000 barrels of oil. Whatever total amount of biogas is available in these urban castoffs, it is certainly significant. It can be recovered through pyrolysis, fermentation, solar energy photosynthetic processes, and perhaps other methods yet to be tried. The logic of processing refuse for its energy and other resources is convincing, whatever the final Btu value. Perfecting methods of accomplishing this efficiently makes both energy and environmental sense. Turning on the lights with today's waste seems infinitely more rational than using it to pollute the nearest stream.

Energy from manure gasification is another practical possibility for which effective and profitable methodology is already available.

> Anaerobic digestion . . . holds greatest promise in the recovery of energy from animal manures because it seems also to be the most economical and yet environmentally satisfactory method of treating such wastes, especially those from confined rearing operations. Manures are readily decomposable and in a state readily amenable to being slurried . . . . The amount of gas produced per unit amount of manure is significantly influenced by the food intake of the animal. Thus gas production per pound of manure from animals raised in the United States is greater than that from animals raised in India.[21]

Agricultural areas are increasingly noting and doing something about the energy opportunities available to them in manure gasification. Era, Inc., for instance, a company in Lubbock, Texas, announced agreements with the Natural Gas Pipeline Company of America to supply biogas (1.2 billion ft^3 annually) from manure gasification. This 1.2 billion ft^3 of biogas is reported sufficient to heat 7000 homes per year.

In the process, manure is converted to methane, with animal feeds and fertilizers as by-products. Cost of the methane is expected to be competitive with nonregulated natural gas. Approximately 200 million tons of manure are produced annually by American agriculture, and it is estimated that up to 15% of U.S. energy needs could be met through biogas from this source. Methane from manure is another example of solar energy reaching us by an exceptionally roundabout route.

MINNESOTA PEAT

Minnesota peat isn't a professional basketball player or a pool shark like Minnesota Fats. It is another solar energy deposit man is planning to withdraw for future use.

Northern Minnesota is one part of the U.S. with extensive peat deposits, and plants to convert this peat to methane have been recommended to relieve natural gas shortages. What is peat? It has been classified as a fossil fuel by some and described as "geologically young coal." It is composed of plants that have undergone a process of decomposition for long periods in areas with plenty of water. Water inhibits aerobic bacteria from breaking down plant materials quickly. Carbon is retained in the cellulose of the plant, and peat slowly develops. Surface peat in Minnesota is about a century old. Peat depths vary from 6 to 20 feet, and the age at the lower levels is estimated at 10,000 years.

Peat is replenished in part every year, with as much as 15 million tons added to the Minnesota supply annually. If just this amount of peat were converted to methane, 75% of the state's natural gas needs would be met. Minnesota is rich in peat, with peat bogs covering areas as vast as 100,000 acres. Ireland is more famous for its peat bogs, but none are larger than 30,000 acres.[22]

Environmentalists criticize plans to mine Minnesota peat. They are afraid of strip mining on a disruptive scale. "For 20 years of methane we're going to wipe out a massive chunk of the earth's surface," warned a forester who prefers considering other alternate energies first.[22] Peat mining could also increase the danger of flooding, though mining advocates insist this is a controllable problem.

Worldwide, peat represents about 1.1% of fossil fuel resources. It has been used as a fuel for centuries in such countries as Denmark, Germany, Ireland, and the Netherlands. Currently the Soviet Union holds the lead as the world's number one consumer of peat fuel (70 million tons in 1975 to produce electricity in 77 power plants). With rising fuel costs, Finland and Greece have also been turning to peat for electrical generation. Most

European applications have used peat in the traditional way, as fuel. The Minnesota plan to process peat with methane as the energy result is unique at present.

Peat resources are chiefly concentrated in the world's temperate zones where the climate for several millennia has suited peat formation. The Soviet Union is thought to have approximately 60% of the world's peat. Finland and Canada account for 20% between them. The U.S., not including Alaska, has perhaps 5%. Minnesota, New York, Florida, Maine, Michigan and Wisconsin have large peat reserves, Minnesota leading with 7.5 million acres. Alaska may dwarf them all. Frozen under the Arctic ice, Alaska could have 50 to 100 million acres of peat.

Will mankind in the future find ways to extract that fuel from below the ice? Oil from deep beneath the Alaskan ice is now being used in the southern 49 states. A safe prediction might be that eventually the frozen peat of Alaska will be converted back again to energy, even if men must cut through the ice. Men have shown willingness to go just about anywhere, cut through just about anything, and do whatever is necessary to have their daily portions of energy. Peat is not a major fuel—yet. Other fossil fuels are still more plentiful and convenient to use. But the others, of course, are exhaustible; and solar energy considerately keeps adding slowly but steadily to the peat supply. Peat, you see, unlike coal and oil, is a youngster in the solar family; and the sun looks after its children, including Minnesota peat.

THE SOLAR CLAN

It's odd, that solar clan, with more members and different ways of behaving than it is simple to keep straight. But starting with old solar energy itself, direct and plentiful, and working through the various members of the family, one fact is plain as tomorrow's sunrise. No doubt it is the oldest and most important fact on earth: The sun supplies all the energy we could ever need or want to use. Solar energy comes to us in so many ways, there seems almost a calculated effort to make it easy for us. But whether that fanciful supposition is true or not, the energy abounds. Most of the work is done for us in advance. About all we have to do is pick up the energy and use it. What more could we want?

THE SOLAR FAMILY - REFERENCES

1. Cottrell, F. *Energy and Society* (New York: McGraw-Hill, 1955).
2. Gimpel, Jean. *The Medieval Machine* (New York: Holt, Rinehart and Winston, 1976).
3. Williams, J. R. *Solar Energy: Technology and Applications, Revised Edition* (Ann Arbor, Michigan: Ann Arbor Science Publishers, 1977).
4. Morse, Frederick H. and Melvin K. Simmons. "Solar Energy," *Annual Review of Energy,* Volume 1 (Palo Alto, California: Annual Reviews Inc., 1976), pp. 131-158.
5. Sutton, O. G. *Micrometeorology* (New York: McGraw-Hill, 1953).
6. Ridgway, James. *The Last Play,* (New York: E. P. Dutton & Co., Inc., 1973).
7. Soucie, Gary. *Audubon,* Vol. 76, May 1974, pp. 81-88.
8. Sørensen, Bent. "Energy and Resources," *Science,* Vol. 189, No. 4199, July 25, 1975, pp. 255-260.
9. *Proceedings of the Solar Sea Power Plant Conference and Workshop,* National Science Foundation (RANN), Sponsor, Carnegie-Mellon University, Pittsburgh, Pennsylvania, June 1973.
10. "Applying Physics to Clean Energy Needs," *Environmental Science and Technology,* Vol. 9, No. 2, February 1975, p. 104.
11. Sullivan, Walter. "A Search for Energy Goes Back to Basics," *The New York Times,* November 3, 1974, p. E9.
12. Dugger, G. L. "Ocean Thermal Energy Conversion," *Solar Energy for Earth,* H. J. Killian, G. L. Dugger and J. Grey, Eds. (New York: Am. Inst. Aeronaut. Astronaut, 1975).
13. Wertenbaker, William. "Mining the Wealth of the Ocean Deep," *The New York Times Magazine,* July 17, 1977, p. 14.
14. Szego, G. C., J. A. Fox and D. R. Eaton. "The Energy Plantation," Paper No. 729168, Proc. IECEC, September 1972, pp. 113-114.
15. Donovan, P. and W. Woodward. "An Assessment of Solar Energy as a National Resource," NSF/NASA Solar Energy Panel, University of Maryland, December 1972.
16. The National Energy Plan, Executive Office of the President, Energy Policy and Planning, April 29, 1977, pp. 75-77.
17. Abelson, Philip H. "Energy from Biomass," *Science,* Vol. 191, No. 4233, March 26, 1976, p. 1221.
18. North, W. J. *Evaluating Oceanic Farming of Seaweeds as Sources of Organics and Energy,* Prog. rep. June 1 to December 1, 1974, under NSF/RANN grant GI-43881 to California Institute of Technology, Pasadena, California. Rep. No. NSF/RANN/SE/GI-43881/PR/75/1, Washington, D.C., National Science Foundation, 1975.

19. Abelson, Philip H. "Energy Alternatives for Brazil," *Science,* Vol. 189, No. 4201, August 8, 1975, p. 417.

20. "Disposing of Solid Wastes by Pyrolysis," *Environmental Science and Technology*, Vol. 9, No. 2, February 1975, p. 98.

21. Golueke, Clarence G. and P. H. McGauhey. "Waste Materials," *Annual Review of Energy*, Volume 1 (Palo Alto, California: Annual Reviews Inc., 1976), pp. 257-277.

22. Boffey, Philip M. "Energy: Plan to Use Peat as Fuel Stirs Concern in Minnesota," *Science,* Vol. 190, No. 4219, December 12, 1975, p. 1067.

CHAPTER 9

GEOTHERMAL, TIDAL AND WATER POWERS

"There is never much in a miracle, when it is explained."

Tide Marks
H. M. Tomlinson

These familiar sources of energy are the same as solar power in that they are continuous and inexhaustible. Unlike solar energy, however, they are less ubiquitous and for the future probably must be relied on as auxiliary sources rather than primary sources of energy. Fusion and solar energy will necessarily serve as the universal fuels of mankind because they have the potential to deliver the vast quantities of energy required. These other energy sources, with smaller though far from insignificant quantities available, will be supplemental. But supplements are important in the human diet, whether food or energy.

GEOTHERMAL POWER

Stand at the edge of an active volcano sometime and you will learn a fact with important energy ramifications. The earth's interior is hot, and thus a vast natural reservoir of heat energy.

This heat can be put to work for man. It can also destroy. In the case of volcanoes, the destruction may be swift and devastating. July 18, 1977 there were news announcements that Etna, an active volcano on the east coast of Sicily, might be close to another eruption. Etna eruptions, killing hundreds of thousands, have been recorded since 475 B.C. The volcano shows no signs of tiring, which supplies proof of the great forces underfoot in the earth's interior. But if these forces can harm, they can also benefit. Parts of Italy may be threatened by volcanic eruptions and the earthquakes that often accompany them, yet Italy also has derived useful energy from volcanic sites.

> Power plants using steam from wells drilled in volcanic areas have been in operation for more than half a century. The first such plant was installed near Larderello in Tuscany, Italy, in 1904. Subsequently, Italian power capacity from geothermal energy has been progressively increased to a present figure of about 400 megawatts.[1]

In volcanic areas where vents or weaknesses exist in the earth's crust, pressure brings heat effects close to the surface where they can be utilized for power. Many parts of the world, enjoying the dubious fortune of being volcanic, are suitably situated to benefit from this source of power, although there is still much to be learned about it and technological problems to solve.

There are approximately 500 active volcanoes in the world today, and a number of other locations with "channels of weakness" for the exodus of geothermal heat. The majority of volcanoes exist in relatively close proximity to the sea, and steam as a symptom of vulcanicity is fairly common. Such steam can be utilized for power. It can also be, apparently, a contributing factor in volcanic eruptions, as in the Krakatoa disaster of 1883 when the steam and ash eruption reached a height of 20 miles, an island split apart, and tens of thousands were killed in the resulting tidal wave. Geothermal power is not a casual, Sunday afternoon picnic. Yet if such power is understood and intelligently managed it can provide valuable supplemental energy.

"Hot spots" or areas with geothermal resources are found in many parts of the world. Hot springs have been used for their believed curative effects at least 2000 years. Hot mud baths at Balaruc-les-Bains, France, are known to have drawn visitors for treatment since early Roman times. Scholarly works have been written on fangotherapie, the French word for this treatment, such as Dr. Astruc's 1737 instruction concerning the time to remove a patient from his thermal therapy.

> From time to time, you inspect the veins on the poor man's forehead. When they're blue and bulging, you remove him from the bath, and put him in a warm bed. If he passes out, you force some wine down his mouth until he's thoroughly intoxicated, then you give him some consommé and carry him away.

In modern Iceland, it is a standard joke about the Icelandic hot springs that "it takes only three minutes to soft boil a tourist." The message is clear that there is heat down there trying to get out. In volcanoes and hot springs, it emerges without man's assistance. In other places, human technology can appropriate the waiting geothermal energy, which quantitatively is great.

> Dr. Don White, of the United States Geological Survey, in Menlo Park, California, estimates that the heat in the top 10 miles of the earth's crust equals 3×10^{26} calories. That is about 2000 times the amount of heat that would be produced if we burned the world's entire supply of coal.[2]

The president of a geothermal energy company in California, a pioneer in the field, believes there is sufficient geothermal potential in the Imperial Valley of California alone to supply the electrical needs of the entire Southwest a minimum of two centuries.[2]

SITES FOR EARTH ENERGY

Despite this potential, geothermal energy is only beginning to attract the funds and effort that will bring improved technology and effective use. In addition to Larderello in Italy, other areas successfully using geothermal energy include Wairakei, New Zealand, and Luzon in the Philippines During the 1970s, steam wells were drilled near active volcanoes on Luzon. Drilled to depths of 6000-9000 feet, the wells are intended to provide steam for up to 5000 kilowatts per well. There are plants generating steam power from geothermal sources in Mexico, Japan, the Soviet Union, and El Salvador. In the U.S. a plant at The Geysers in northern California earned this report in 1974:

> Most of the talk about geothermal energy since the advent of the energy crisis has focused on its electricity-generating capability. Indeed, the largest commercial geothermal producer of electricity in the world is the Geysers Power Plant of the Pacific Gas and Electricity Company in Sonoma County, California. By the end of 1976 this facility will be generating 908,000 kilowatts of electricity—enough to serve a city the size of San Francisco, 90 miles away.[3]

The Geysers is located in what geologists call the "ring of fire." This is an area where geothermal energy is often available because "channels of weakness" in the earth allow molten magma to rise through faults in the earth's crust. Magma is the source of heat energy for steam wells or other methods of exploitation. The "ring of fire" runs through coastal areas of the Pacific, including Alaska, the western U.S., Japan, the Philippines and New Zealand. Along this ring are potential producers of geothermal energy from wet or dry steam.

At The Geysers and Larderello, dry steam is used to generate electricity in a process that is simple and direct. Other places use hot water as the energy source, and the technology is more complex. At Cerro Prieto, Mexico, for instance, a 76-megawatt plant uses a flashed-steam process. Hot water rises from the earth, undergoes pressure reduction, "flashes" into steam, and then is used to operate a generator.

A variant method is not to let the hot water flash into steam, but to keep it pressurized. The hot fluid is pumped into a heat exchanger where its heat vaporizes isobutane or Freon, which in turn operates a turbine. In the never-flagging pursuit of greater efficiency for energy utilization, the Lawrence Livermore Laboratory is designing a "total flow system" to use both steam and hot liquids for power conversion.

> In principle, it is related to the impulse turbines that have been used in hydroelectric power plants for more than a century. Water from a dam or some other source is expanded through a spray nozzle; the force of the droplets acts against the scoop-like vanes of the turbine wheel, which then turns to operate a generator.[2]

California's Imperial Valley lies over areas of hot brine sufficiently close to the surface for energy plants to exploit. Enormous amounts of heat are available in this brine. There may be even greater heat in dry-rock formations throughout the western U.S. An area of 95,000 square miles, including all or part of 13 western states, has been estimated to have,at depths of about 3.5 miles, hot dry rocks with temperatures averaging 288°C (550°F). Techniques are available and being improved to use this heat. Effective new technology in the geothermal energy field has an excellent potential for a good energy return. The 1977 National Energy Plan says of this source:

> Hydrothermal (liquid-dominated) sites are found throughout the West, some at high temperatures adequate for electricity generation, and others at lower temperatures suitable for heating of buildings. At present, several hundred buildings use geothermal heat. With expected technological progress, hydrothermal sources should begin to make a significant contribution in the 1980s. Geopressurized resources, located along the Gulf Coast, contain potentially significant amounts of hot water and dissolved methane, which may become accessible in the 1980s. Hot dry rock may become a significant energy source in the 1990s.[4]

DRY GEOTHERMAL METHODS

A promising way to utilize geothermal energy from hot underground rocks involves drilling into the rocks and injecting water from the surface. The water changes to steam and is recirculated to the surface for energy use. Research efforts are going forward at numerous western locations. At Los Alamos, New Mexico, a government laboratory is undertaking a special geothermal variation. The plan is to bore two holes in the hot rock, then crack the rock with fluid under pressure. Water sent through the cracks is expected to return to the surface as steam, ready for use in a power plant. One objection to this method has been that water piped down, because of its pressure and temperature, would eventually close the cracks and stop the steam flow. Los Alamos scientists will seek to counter this objection with innovative technology. In developing most energies, fast technological footwork is a prime requisite.

157

U.S. "HOT SPOTS" ARE PLENTIFUL

The U.S. Geological Survey has identified 1.8 million acres in the U.S. as Known Geothermal Resource Areas. Another 99 million acres are thought to have geothermal energy potential. Applying currently available technology, a minimum of 12,000 megawatts of electric generating capacity is available from geothermal sources (15 times the 1975 geothermal usage rate). Geological surveys show that considerably greater geothermal energy is immediately available for space heating.[5]

California and New Mexico are the best known prospects for U.S. geothermal development. They are the "high-temperature resource" areas. Intermediate-temperature areas are widely scattered and can also supply useful energy. Next to be developed are geopressure zones, such as those in Texas and Louisiana, which contain hot water at high pressure. Geopressure zones can supply energy two ways: (1) heat is withdrawn from the water; (2) the high pressure is used as a mechanical energy source.

AMPLE ENERGY, BUT . . .

In 1977 geothermal energy, the same as other alternate energies,was available in large quantities, but the price was too high. The story is familiar. The energy is there. Ways of using it effectively already exist. Even better ways are certain to be found. But at the present time, with the exception of The Geysers and a few other locations, most geothermal sites could not supply electricity at competitive prices. Geothermal makes one more promising and waiting energy for the "not yet economical" list. Its existence as a permanent option should be reassuring. Could it also be responsible for dangerous complacency and apathy among Americans, leading them to use up fossil fuels at a breakneck pace, to shrug off pleas for conservation, and to do so under the assumption that the "not yet economicals" will keep them going when the time comes? The problem with this assumption has been the tendency to go slow in developing technology that will enable geothermal and the other alternates to perform adequately. If a sense of urgency doesn't prevail a long time before the alternates must be brought on line, will the technology be ready to make use of alternate energy resources? These complicated questions were at the heart of the U.S. energy crisis in the 1970s.

GEOTHERMAL ENERGY AND THE ENVIRONMENT

Environmentalists consider geothermal power a potentially inexpensive and convenient source of energy for many areas, but they are critical of claims that call it a "clean" energy source, noting such problems as noise pollution in the vicinity of geothermal power developments, release of uncomfortable and perhaps dangerous odors, and the disposition of mineral-laden runoff waters.[6]

The point is also made that although geothermal power may be inexhaustible, the plants using it at particular sites will not function permanently. Each plant will draw off the thermal energy stored in its area, for periods of possibly half a century, and then be inoperative.[7]

Additional drawbacks include the possibility of land subsidence as a result of withdrawing fluids from geological formations (earthquakes also conceivably could be stimulated by such action), and the potentially more serious threat of radioactivity at geothermal plants. Various studies in Maine, Great Britain, and Austria have tended to confirm this. In an analysis of this problem, the following was noted:

> Any nongaseous radionuclides present in the discharge (such as ^{226}Ra) will enter the river. In addition to environmental releases, radon daughters may build up on interior surfaces of power plant equipment in a situation analogous to that recently discovered in natural gas processing. This potential buildup could result in occupational exposure to radiation and radioactivity.[8]

These are speculative concerns, but welcome. Sharp consideration of possible environmental hazards in advance will help prevent the mistakes of the past—rushing into a new technology

159

without assessing its side effects—and will alert us to find solutions before problems swallow us whole. That is why the increasing attention paid by energy developers to environmental matters may be costly in the short run, but is likely to be the far better course, and even the more economical course in the long run. Maintaining a reasonably habitable place in which to use whatever energy we have is beginning to be accepted as a practice that makes sense.

Geothermal energy appears to make very good sense. Despite some unsettled environmental issues, geothermal energy is essentially considered one of the "benigns," with no critical environmental hazards that can't be handled. It is conjectured that geothermal energy could be employed successfully in the production of electricity, water heating, space heating, refrigeration, air conditioning, and the delivery of process steam. Estimates indicate as much as 20% of U.S. energy needs might be geothermal in origin, given expected technological advances and competitive prices. When this will happen depends on other energies and their costs. Future energies, the same as past energies, will inevitably compete among themselves, with the consumer's funds going to what seems the best buy. At the moment, geothermal energy is not the best buy; but the future is open-ended and conditions alter.

Geothermal energy exists, it probably will be needed, and in some places is waiting to be used. With world energy demands rapidly multiplying, future energy needs will not encourage observing a reasonably accessible source go to waste. Even as the sun dispatches energy across 92 million miles of space to the earth's crust, so the earth itself brings energy from within to the surface. Energy is provided. Man's task is to find the best way of using it.

TIDAL POWER

Four times each day the phenomenon occurs. The moon conducts its 24-hour and 50-minute rotation about the earth, and in response to the rotation, two high and two low tides appear.

The tides have been so predictable, so inevitable, they have always been part of man's folklore and superstition. There are probably still people in coastal places who believe as did Mr. Peggotty in Dickens' *David Copperfield* that people die only when the tide goes out, and are born only when the tide comes in.

Through most of man's history, poets, not scientists, have often been the principal students of the tides. The poets have viewed this evidence of unalterable gravitational attraction between earth and moon as a supreme symbol of eternal regularity and necessity.

> " . . . Like as the waves make toward the pebbled shore,
> So do our moments hasten to their end."
>
> (Shakespeare)

> " . . . I must go down to the seas again, for the call of the
>   running tide
> Is a wild call and a clear call that may not be denied."
>
> (Masefield)

> " . . . Sophocles long ago
> Heard it on the Aegean, and it brought
> Into his mind the turbid ebb and flow
> Of human misery, we
> Find also in the sound a thought
> Hearing it by this distant northern sea."
>
> (Arnold)

Philosophical poets have not been the only ones watching the tides go out and return, however. Other men have observed the diurnal motions to and fro with thoughts focusing not on eternal rhythms but on what could be done with all that perpetually repeated energy. The use of tides for power is by no means a discovery of the twentieth century. It could even be argued that contemporary efforts to use the energy of tides is another example of today finally catching up with the twelfth century.

ENERGY LESSON FROM THE MIDDLE AGES

As early as the twelfth century and increasingly throughout the Middle Ages, tides were used by Europeans to operate mills. When the tide flowed in, water was held in ponds that contained dams with swinging gates. When the tide ebbed, the gates would close from the pressure of the water. The tidal waters thus dammed could then be used to operate the waterwheels of a mill.

Jean Gimpel in *The Medieval Machine* writes that tidal mills "are typical of the medieval urge to discover new sources of energy." He pays tribute to the astuteness of medieval engineers:

> So remarkable was their choice of sites for tidal mills that the first twentieth century tidal-powered plant, built after World War II by the Electricité de France . . . dammed a river along which lay a whole series of medieval tidal mills which were still in active use, on La Rance, near Saint-Malo in Brittany.[9]

If the technological awareness and drive of the 12th to the 14th centuries had been maintained in later centuries, we might be farther along today in effective use of tides to help supply human energy needs. But no harm done. The complacent adage of Livy comfortably applies to the utilization of constantly renewable energies: Potius sero quam nunquam (Rather late than never). In fact, if we had put off using fossil fuels until now, think of all the oil we could have. Yet even so, the grand rush would no doubt start to burn it as quickly as possible, and soon enough we should be right where we are, looking gratefully and hopefully at the tides and other constant energies. Indeed, we might be repeating Shakespeare's sixteenth century words and recognizing them as prophetic for energy development in the twentieth century: "There is a tide in the affairs of men which, taken at the flood, leads on to fortune."

Unless men do something to eliminate the moon, or the oceans, tides will continue to make their rounds inexhaustibly. We can count on that, and on the fact that great amounts of energy are ours for the taking. Certainly full exploitation of the tides can provide man with a tidy and much-needed supplementary fortune in energy. Electricité de France took the lead with its postwar development of a tidal energy facility. The first large-size plant designed to use tidal power as an energy source began operating in 1966 at France's La Rance estuary. The plant was designed with a capacity of 320 megawatts.[1] The same author discusses Russian plans for tidal power plants and notes that large plants are feasible only in a few locations. Required are a "combination of a large tidal range and a bay or estuary capable of being enclosed by dams."[1] When conditions are right, however, the energy return is substantial.

BAY OF FUNDY

To appreciate the scope of the power involved, consider the Bay of Fundy. Between the Canadian Provinces of New Brunswick and Nova Scotia, the Bay of Fundy is 148 miles long and 48 miles wide at its mouth. The rise and fall of tide at the Bay of Fundy is one of nature's remarkable displays. At the head of the Bay, tides of 50 feet are frequent, and they have sometimes exceeded 60 feet. Tidal power in the Bay of Fundy alone could average 29,000 megawatts.[10]

Even if the use of tidal power is necessarily restricted to favored places such as the Bay of Fundy, potentially more than 60,000 megawatts are available from this source. The energy can be obtained by means of tidal basins (hence the dams) and hydraulic turbines. There is little environmental upset, and the energy reaped from the sea as the moon passes is "clean."

When future energies are developed, it would seem less than fully responsible not to take the tide at its flood and allow ourselves to be led on to whatever fortune in energy is available.

WATER POWER

Anyone who, while walking on the street, has been hit suddenly by a stream of water from a second floor apartment can testify to the energetic force. "Gardyloo!" (*gardez l'eau,* beware the water) was the word housewives of old Edinburgh used to cry in warning before emptying water, etc., into the streets.

Drenched citizens of Edinburgh who didn't gardyloo in time would certainly have had a few words to say on water power.

Quite a lot, in fact, can be said. According to Hubbert, approximately 210,000 megawatts of water power capacity were utilized worldwide during 1964. That amount was only 7.5% of the overall potential.

Water power has been used by man to do work since antiquity. Records of water used to turn waterwheels in grinding corn and other human labors have existed many centuries. Antipater, a poet in the time of the early Roman emperor Augustus, described this energy in action:

> Down on the top of the wheel,
>   the spirits of the water are leaping,
> Turning the axle and with it
>   the spokes of the wheel that is whirling
> Therewith spinning the heavy and hollow
>   Nisyrian millstones.[11]

Prior to the use of water power, mill wheels were laboriously turned by human labor, often slaves. Applying water to turn such wheels was one of the early instances of freeing men from burdensome ordeals through the utilization of energy. Water power was used in Europe from before the beginning of the Roman Empire into modern times; and as the use of water power increased, slavery declined. Thus the broadening availability of energy other than human energy helped eliminate the need for slaves and to bring about the gradual disappearance of the practice. In time the use of horses and oxen as beasts of burden declined as well for the same reason. Other energy sources could do their work better and more economically. The moral betterment of man is one clear dividend of advances in energy technology. There is also an implied warning. If plentiful supplies of nonhuman energies are not maintained, will the peculiar institution of slavery grow again in popularity? It would be nice, but naive, to pretend that the answer is no. Fortunately, with undeveloped water power and other future energies available, that difficult question hopefully will never be asked in earnest. Or will it?

Thousands of waterwheels served European agricultural and industrial (*e.g.,* papermaking, cloth production) needs through the Middle Ages. Wherever it was available, water power attracted users in Europe and later in America straight up to the present.

HYDROELECTRIC POTENTIAL

Hydroelectricity is the chief modern goal of water power development. In this application, water power offers an enormous energy source for many parts of the world. Africa and South America, for instance, have the world's largest water power capacities of 780,000 and 577,000 megawatts, and both continents lack significant coal deposits.

Hydroelectricity is another gift from our old friend, solar energy. The sun's power circulates water from oceans to high elevations through evaporation and then precipitation as rain or snow. The water finally works its way via rivers and streams back to the sea, completing what is called the hydrologic cycle. As the water flows from high elevations to sea level, it delivers energy that can be translated into electricity. It also contains energy that can be translated into disaster in the form of floods.

So there are problems. The criticism is made that the large dams and associated activities necessary to achieve maximum electric power also promote conditions favoring maximum flood damage. If the world's total amount of water power were put on the line, it would equal all the energy now being consumed. But the drawbacks cannot be dismissed. Most potential water power sites require large dams to produce extensive reservoirs. And there are many, including the President of the U.S. in 1977, who do not love big dams.

Hydroelectric power is obtained by constructing a dam and installing a turbine plus generators in the base. Water is held up by the dam, and when it is allowed to pass through in controlled flow, it operates the turbine much as it operated ancient waterwheels to grind corn. In this case, the result is not flour, but electricity. Great pressure is achieved by water as it builds up hundreds of feet high in some dams. When released, intense force acts on the turbine, with large amounts of electricity produced. In the Pacific Northwest of the U.S., hydroelectricity (chiefly from the Columbia River) has provided over 90% of the electrical power used in the region, with electricity costs substantially lower than in other parts of the country.

In 1969 the capacity of U.S. hydroelectric developments was 53 megawatts, and the amount of energy produced was 17% of U.S. electricity, or 4% of the nation's total energy use.[1][2] In the 1977 National Energy Plan, expanding hydroelectric capacity was emphasized:

> New or additional hydroelectric generating capacity at existing dams could be installed at less than the cost of equivalent new coal or nuclear capacity. Many of these sites are small, but could generate 3 to 5 megawatts, and are located near major demand centers currently dependent on imported fuel oil. Installation of additional generating capacity at existing sites could conceivably add as much as 14,000 megawatts to the nation's generating potential.[4]

Where water power naturally exists, it is an undisguised boon. At the start of the seventies in the U.S., 45,000 megawatts were being used from a total estimated capacity of 161,000 megawatts.[1] The Pacific Northwest and Alaska are the sections of America with the best locations for further development of hydroelectric power. The Federal Power Commission reported 125+ megawatts as the wattage still undeveloped. In the 1970s, it was uncertain how much of this potential would be achieved. Water development projects were cancelled in 1977 when the Executive branch of the U.S. government found them unnecessary and environmentally harmful.

BIG DAM CONTROVERSY

The environmental difficulties associated with water power are familiar. Big dams are the objectives of those who want maximum power from water, yet big dams are often environmentally calamitous, unless park lands and national monuments are preferred under water. Water power and big dam devotees have been gazing avariciously at the Grand Canyon in Arizona for many years. Some of these devotees aren't convinced by the argument that it is a bad trade to swap the natural majesty of the Canyon for the power to operate Channel Four at the flip of a switch.

Seepage from reservoirs has raised water levels in surrounding soil and at the same time raised mineral levels that decrease soil fertility and productivity. Earthquakes, according to some theorists, may result from the heavy weight of water in large reservoirs straining the earth's crust. These are environmental critiques. On the energy-efficiency side is the tendency of reservoirs over a period of time to fill with sediment, gradually lessening electrical output. Preventing reservoirs from silting requires persistent supervisory care to assure optimal long-term performance.

As with other immense prospective future energies, in connection with water power, responsibility is the keynote. Irresponsible pursuit of power for its own sake may end by ravaging the land, leaving it defenseless against floods. The wise use of water power requires long-range conservation policies based on the intimate relationship between the land and the water. Without such policies, the result could be another case of an energy orgy now and starvation later.

A responsible balance is essential between energy needs and the ecological, geographical needs of an area in which hydroelectric developments are possible. Careful planning is an obligation the present owes the future whenever schemes are afoot to tamper with the normal flow of a river. Objective studies and assessments must be made to avoid doing more harm than good, to supply power while keeping the environmental impact to a practical and acceptable minimum. Reasonable men will not blindly insist that there must·be no dams whatever, that we must try to keep things exactly as they are (something nature never tries, incidentally, and actively opposes with a familiar pattern of constant change). Energy advocates when reasonable will not demand "More hydroelectricity, whatever must be destroyed to get it." The challenge is persuading men to be fair and to recognize that the best answer in a judicious pursuit of energy typically is a compromise seeking to reconcile both sides of the question.

Since much of the earth's water power is yet to be developed, perhaps both earth and energy wisdoms can be allowed to prevail. When developing water power or other potential energies, it is logical to be reminded that this earth is man's permanent habitation. With that truth in view, perhaps the prudence and care men give their houses can also be granted the earth. All of us are earthowners, of course, with corresponding responsibilities.

GEOTHERMAL, TIDAL AND WATER POWERS - REFERENCES

1. Hubbert, M. K. "Industrial Energy Resources," *Nuclear Power and the Public,* H. Foreman, Ed. (Minneapolis, Minnesota: University of Minnesota Press, 1970), pp. 179-206.
2. Henahan, John F. *Popular Science,* Vol. 205, November 1974, pp. 96-99.
3. Wright, Robert A. "Klamath Falls, Oregon—Living in Hot Water," *The New York Times,* October 13, 1974, p. F3.
4. The National Energy Plan, Executive Office of the President, Energy Policy and Planning, April 29, 1977, p. 73, pp. 77-78.
5. White, D. E. and D. L. Williams, Eds. "Assessment of Geothermal Resources of the United States—1975," Geological Survey Circular 726, Washington, D.C., 1975.
6. *The Denver Post,* "Environmentalist's View," June 12, 1974, p. 36.
7. White, D. E., *Geothermal Energy,* U.S. Geological Survey Circular 519, Washington, D.C., 1965.
8. Gesell, Thomas F. and John A. S. Adams. "Geothermal Power Plants: Environmental Impact," *Science,* Vol. 189, No. 4200, August 1, 1975, p. 328.
9. Gimpel, Jean. *The Medieval Machine* (New York: Holt, Rinehart and Winston, 1976).
10. Trendholm, N. W. "Canada's Wasting Asset, Tidal Power," *Electrical News and Engineering,* 70, 1961, pp. 52-55.
11. de Camp, L. Sprague. *The Ancient Engineers* (Cambridge, Massachusetts: M.I.T. Press, 1963).
12. Holdren, John and Philip Herrera. *Energy,* Sierra Club, 1971, pp. 52-56.

CHAPTER 10

ASPECTS OF THE FUTURE

"There are two principles that have been cornerstones of the structure of modern science. The first—that matter can be neither created nor destroyed but only altered in form . . . . The second—that energy can be neither created nor destroyed but only altered in form—emerged in the nineteenth century and has ever since been the plague of inventors of perpetual-motion machines; it is known as the law of conservation of energy."

Henry D. Smyth, 1945

"For myself, it is true; I know no care at all. But the faintest disturbance of equilibrium is felt throughout the solar system, and I feel sure that our power over energy has now reached a point where it just sensibly effects the old adjustment. It is mathematically certain to me that another thirty years of energy-development at the rate of the last century, must reach an *impasse*."

Henry Adams, 1902

CONFRONTATION

When New York City's lights went out on a summer night in July 1977, we were grimly reminded of the fact that our civilization is in trouble and that it is dangerously, almost pathologically vulnerable. One of the major cities of the world was rendered impotent and helpless simply because electricity was temporarily unavailable. Elevators would not operate. Water could not be delivered to apartments in high-rise buildings because pumps were shut down. At hotels using electronic security systems to open room doors, hotel guests could not enter their rooms because the electricity was off. Some New Yorkers rose to the challenge with grace and good humor. Thousands of variations were delivered on the joke that the lights would go on again when the city paid its bill. But there was tragedy as well as nuisance, in fact mostly tragedy as a catastrophic and uncontrollable binge of destruction swept the city. Poor people, driven by hate and confused by hopelessness, rose up in the darkness and destroyed their own neighborhoods in an orgy of demolition and theft.

Everything that happened can be traced to common roots, where we find the failures of our energy systems and our social systems to protect modern man from the consequences of his own excesses and dependence. Without electricity, his food spoiled, he had no water, and a gang of youths was robbing him. Physicist Amory Lovins wrote about energy after the New York debacle:

> A transition to a benign, resilient, and sustainable energy system will take a long time—perhaps 50 years—but if we do not start soon, the option will slip away, foreclosed by other commitments. The alternative—the ever-more-electrified, centralized, large-scale and vulnerable system that we have been unthinkingly weaving into our lives—would insure ever larger, more frequent, and less reparable failures. That choice is framed for us by the New York blackout. If we see the blackout as a sort of warning "heart attack," signaling the need to reexamine our approach to the energy problem and the myriad other problems intertwined with it, then we may live to be grateful for the shock.[1]

Lovins isn't a lonely voice crying in the wilderness. He has plenty of company. It is now clear to most that we are living through a time of transition, passing from an era of plenty to an era of . . . there's the rub. We can't be entirely certain what is ahead, whether an era of painful insufficiency, or just barely enough, or what. Will Rogers nearly half a century ago wrote, "We are going at top speed and we are using our natural resources as fast as we can . . . But when our resources run out, if we can still be ahead of other nations, then will be the time to brag; then we can show whether we are really superior."

Perhaps we are outgrowing an adolescent impulse to travel at top speed and to show ourselves superior. Perhaps there are subtle benefits for us in learning to cope with the dilemma of an abundance coming to an end. Then we may learn that abun–dance really doesn't come to an end at all, but is *shifted* from one avenue to another. Indeed, we may through the pressure of necessity and our reaction to it, make the time ahead an era of intelligence. Stranger things have happened, though admittedly not much stranger.

> In energy . . . good prospects exist if we possess the will and patience to conserve what we have while alternate energies are developed. New technology in resource use, energy con-servation, and material recycling doesn't guarantee Utopia, but new technology does support the hope that mankind can, with sufficient and consistent effort, avoid losing control. Men still have the capacity and the tools to save themselves.[2]

QUESTIONS

Men in every age are challenged by nature, and their reponse to that challenge becomes the critical measure of their human achievements. In America we may be learning that the most effective challenge might be an intelligent question rather than tyrannical force or abuse. In a classic work over a generation ago, J. W. N. Sullivan wrote as follows:

> The art of putting correct questions to nature is learned only gradually, and there can be no doubt that some of our present difficulties arise from the fact that this art is not yet complete-ly mastered.[3]

Probably this particular art will never be completely mastered, but there is something more important than mastery, and that is to persist in asking questions, right or wrong. The correct ones will prove themselves in time. The incorrect questions will wither away because they aren't watered and nourished by the truth.

It is the peculiar obligation, challenge and opportunity of science to question nature. Asking questions is nearly always the indispensable first step. And when nature is not in a mood to be questioned, when she is secretive and uncooperative, then the interrogators must be particularly insistent.

An essential corollary of asking the proper questions is having the nerve to go on, to resist the temptation to stop too soon, to keep moving forward into the heart of whatever mystery is ahead. The science writer Arthur Clarke in this connection has said, "Let us not be victims of a failure of imagination."

The purest instincts of science are to persist in asking questions. And imagination surely is one of the methods of science in deciding what questions to ask. Supplying the energy needs of man has always required prodigious questioning of nature together with imagination which might also be called creative spunk. The future sciences of energy certainly will not be exceptions. Nature will need the most intensive sort of interrogation, plus the courage to follow where imagination leads.

When fusion energy comes to benefit man, perhaps to free him from the spectral phantoms of insufficient power, behind the accomplishment will be a long record of scientific questioning and tenacity in pursuing truth.

When solar energy is fully utilized to supply man with electricity, heat and cooling as well as traditional vegetables and summer tans, it will be the result of a long questioning process during which scientists and engineers never stopped asking nature "how?" and "why?"

Solar energy and fusion are not the only future energies man will want in his cupboard. Others too may be needed, and they will be available to the extent that man's capacity for questioning and imagination are applied. ERDA has indicated that we can't rely on a single technology to meet our energy needs. Thus we need to nurture all our options. In the year 2020, ERDA expects the U.S. to obtain its energy from coal, nuclear, solar, petroleum, natural gas, hydropower, and geothermal sources.[4] Energy may be sufficient in 2020 if we are smart now, less than half a century from that rendezvous. But it can't be sufficient if we neglect our homework in both research to perfect new energies and energy preservation to conserve and prolong what we have.

ENERGY CONSERVATION AND THE FUTURE

"Must we keep up this tiresome conservation business even in the future?" You probably want to ask that question and season it with a little exasperation. Then add this instruction: "Kindly pick future energies that will supply not only our energy needs but also our energy appetites. What's the future for if not to be better than the present?"

Alas, my friend, that future we are heading for may never again contain an era in which energy is both cheap and plentiful. All the signs are that our golden days of cheap, quick energy are behind us for the most part. None of the future energies available to us are technically simple, and the cost of technology to bring them to our doors will be substantial. Be prepared to pay for your needs and appetites. The amount you will be asked to pay may even convince you it makes sense to moderate your energy appetites. You may decide to be a little colder in the winter, a little warmer in the summer. You may decide to drive less and walk more. You may decide pilot lights on stoves aren't all that superior to matches.

This too should be kept in mind. That future out there will contain much larger populations than the earth now houses. Those populations will first have to be fed. It takes energy, lots of it, to grow food. Then the basic energy needs of the earth's people will have to be met. Whatever future energies we settle on probably will not leave us significant surpluses, certainly not inexpensive surpluses.

So the answer, though it may seem a regrettable nuisance and may cool your enthusiasm for the future altogether, is that something we might as well call conservation will be essential through all the future we can foresee. Only a complete decline and disappearance of civilization as we know it would substantially change that prospect. You needn't give up on the cleverness of technology. You can hope some smart gimmick will make solar or fusion energy so easy and inexpensive to deliver, we can have all the energy we want at prices on a par with petroleum in the depression years of the 1930s. Please do hope—hoping is healthy. But don't count on it too greatly to happen soon.

For a long time we should realistically expect that there will be a sum total of reasonably priced energy available that is somewhat less than we should like to have. The growing energy needs of a growing population make that inevitable. Also inevitable will be the necessity of continuing the tiresome conservation business. This necessity may be permanent, at least until the "smart gimmick" shows up that we can hope for but can't count on. Conservation, of course, is not one of the future energies; though it is axiomatic that if you have energy available, and don't burn it up, you still have it. Translating that axiom into a prudent life style and energy consumption behavior pattern will necessitate sensible conservation to preserve future energies. At a minimum, the human race has no choice but to apply to its energy income the maxim of Mr. Micawber in *David Copperfield:*

> Annual income twenty pounds, annual expenditure nineteen nineteen six, result happiness. Annual income twenty pounds, annual expenditure twenty pounds ought and six, result misery.

Economists may quarrel with this, point to governments as examples, and argue the logic of deficit spending. The question Mr. Micawber might ask those economists and governments is this: Are they happy?

SAVING HABITS

In connection with future energies as well as present energies, all questions land on the shores of one solid fact in the midst of time's ocean: We can only use energy we have, we can only use energy we can afford. There is simply no satisfactory way of applying deficit spending to a ton of coal. It's there or it isn't. It will burn or it won't. You have it or you don't.

Repeatedly we have heard it said that the era of cheap energy is over. The era of surplus energy may also be ending, at least for a time. Whatever future energies we have available are likely to be actively complemented by conservation in its multitude of forms: Eliminating energy waste; exercising temperance

174

rather than gluttony; thinking before we turn on the lights or turn up the thermostat; buying a bicycle or a new pair of shoes. Conservation has many faces, and some of them aren't so gruesome. Collectively the various forms of conservation mean restraint, not hardship, and prudence, not starvation. But it may well be that without self-discipline and the intelligent use of energy, in short without some measure of conservation, the result is both energy hardship and starvation on a broad and devastating scale.

Denis Hayes describes conservation as "the least expensive, most reliable, safest, and least polluting source of energy we can tap."[5] Over half the energy consumed in the U.S. has been attributed to waste.[6] California's environment-oriented governor, Jerry Brown, said in 1976, "I think we need a very substantial commitment to conservation . . . . The biosphere is a very delicate thing. The air, the water, the land, these are limited things; and we have to learn not only to respect them but to reverence them. And that means conservation. Conservation by the kind of automobiles we build, by the emphasis we put on mass transit, by the water we use for flushing purposes in industry, by the manner of lighting that we use, by the whole way that we produce our goods and live our lives. Conservation has to be first." When the governor was asked if he believed this austerity would be accepted, his reply was unequivocating: "I don't think we have any choice."

In the 1977 National Energy Plan, America's necessary future commitment to conservation is emphasized with even greater intensity, and a near missionary zeal:

> The cornerstone of the National Energy Plan is conservation, the cleanest and cheapest source of new energy supply. Wasted energy—in cars, homes, commercial buildings and factories—is greater than the total amount of oil imports . . . America needs to embrace the conservation ethic. The attitudes and habits developed during the era of abundant, cheap energy are no longer appropriate in an era of declining supplies of America's predominant energy sources. Conservation offers vast opportunities for American creativity and know-how. The challenge of saving energy should galvanize the ingenuity and talents of the American people.[7]

175

How effective economic, patriotic, moral and pragmatic appeals for prudent energy use will be is not yet known. The returns from early precincts haven't been encouraging. But we need to remember that our energy addiction may be the toughest to cure because it is the easiest to develop. And it is ridiculously simple to rationalize why one would be foolish to conserve and suffer. Energy consumers decide to believe that there is no energy shortage, that it is all a mammoth hoax to hike prices. Businessmen and industrialists decide to believe that conservation is a hoax to hamper their growth and destroy economic balance. Instilling the conservation ethic requires gently pointing out the conspicuous flaws in these arguments. There is certainly an energy shortage if people cannot buy all the energy they need at a price they can afford. And nothing will suffer more than the economic balance, health and stability of the country if fuel runs short in the midst of business rationalization exercises.

Both government officials and scientists in 1977 seem to agree there is a need for realistic energy conservation to help us through the coming decades. John F. O'Leary, head of the U.S. Federal Energy Administration stated the matter succinctly: "Without an all-out conservation effort, we'll fall off the cliff in the early 1980s."

Of course, "conservation" is simply a word with a variety of complex meanings and interpretations when applied to energy economics, efficiency and social distribution. Applying conservation in specific, energy-saving ways is both a technological and a human problem. Reaching for solutions to the problem will be a familiar calisthenic as we sort out the energies of the future. It might be wise not to put off conversations on the subject until we have a chance for a brief talk at cliffside in the 1980s.

Clearly there are many things we can do individually and collectively, and most are not especially difficult. For one thing we can adopt that conservation ethic they talk about, and give it some teeth and a touch of meaning. There are a hundred things we can do . . . you can do. Name ten. Oh sure you can!

RECYCLING—ONE THING TO DO

In many parts of the world, the challenges of survival have taught people the necessity of letting nothing go to waste. American travellers in parts of the Orient have been struck by the tenacity of people in clinging to anything that may serve a future need. American sailors, including this writer, visiting Hong Kong were both shocked and impressed by the eagerness of a lady we called "Hong Kong Mary" to contract for the ship's food refuse. What we threw away was obtained, cooked in highly seasoned pastries, and sold to the poor of that crowded refugee city as "honeybuns."

Human hunger gave writer Aldous Huxley a stark illustration in India of the "waste not, want not" instinct. Huxley was riding an elephant through a village, when the giant mammal was suddenly inspired to relieve itself of internal burdens in a perfectly natural manner. The result was impressively proportional to the size of the animal. When this occurred, an old peasant woman joyfully ran from her nearby hut with a ready container to take possession of the elephant dung. Dried, it would serve her family as a valuable cooking fuel and help them in the ceaseless struggle to stay alive.

Vance Packard in *The Waste Makers* offered a revealing comparison between Americans and Orientals when he wrote:

> The average American family throws away about 750 metal cans each year. In the Orient, a family lucky enough to gain possession of a metal can treasures it and puts it to work in some way, if only as a flower pot.

Perhaps in an era of growing shortages involving both energy and resources, the American family too will learn the value and logic of not wasting the potentially useful, not discarding items such as cans, papers, and other items that can be recycled.

In many American cities today there is something that did not exist a few years ago: A recycling center. The very existence of such a center is an important departure from the indifference and the waste habits of the recent past. At recycling centers, metal containers, newspapers and magazines, and glass items are

177

welcome. According to Ted Marlin, "the skyrocketing cost of newsprint has made this item the most profitable one for these centers, enabling them to stay in business."[8]

Aluminum recycling has been given national publicity by aluminum companies and represents one of the more successful efforts to use valuable materials over again. Recycling aluminum, writes Marlin, gives a "95% energy saving over the process of extracting the new metal from ore."[8]

Essentially, the recycling movement in America is just beginning. It is an important form of energy conservation, relatively easy to practice, and rewards practitioners. Recycling is likely to spread and expand among individuals, and in other ways too. Scientists are busy seeking ways to extend the benefits of recycling.

One group of scientists at M.I.T. has reported an effective way to convert sewage sludge into an important soil additive. In a sense, they are introducing modern technology to an ancient custom in frugal parts of the world where refuse is collected routinely for fertilizer applications. M.I.T. research indicates that by directing electron beams at sewage sludge, bacteria, viruses, insects, and some poisonous chemicals are rendered harmless. Electron beam treatment was carried out experimentally at a Boston sewage treatment plant and proved considerably more economical than the competitive method of heating sludge to disinfect it. The recycled sludge can be used for soil building. Disinfecting sewage and putting it to work for beneficent purposes is illustrative of the recycling urge in action. Recycling deserves encouragement and imitation. Americans may not come to the point of following elephants (there are so few in America), but they may learn to follow the good practices of their neighbors and save newspapers, cans and bottles for their local recycling center. Old habits are slow to change. New habits are not easy to master. Yet both are certainly achievable. Since waste is a singular folly in a time of energy and resource scarcity, and since recycling is an obvious and painless way to help, practicing it is a new habit that should catch on if the reasons and the methods are publicized and explained.

PERMANENT INVITATION TO TECHNOLOGY

A paradox of the 1970s has been a growing criticism of technology combined with a contradictory assumption that somehow or other technology will eventually come through with the "answer." There is a tendency now to blame technology for the energy woes and other aging pains in our civilization. A case can be made for this argument, though it is weak in logic and tends to wobble. The case goes like this: If technology hadn't put that convenient, fast and ever-ready automobile (when we remember to charge the battery) in the driveway, we wouldn't have been able to drive it around the corner to the butcher's, the baker's, and the candlestick maker's. We'd have walked, been healthier, saved our reserves of energy, and conditions would now be altogether brighter. Technology's to blame.

Swell. Let's all drive to the nearest convention hall and have a grand, old-time festival blaming technology, adding of course that since technology got us in this fix, it is up to technology to get us out, which would be a neat instance of assigning the sickness to cure itself.

Technology, of course, didn't get us in a fix. We got us in a fix by accepting the gifts of technology, such as energy, and using them frantically with blatant disregard for tomorrow's inevitable hangover. So now it seems we may have come to the "morning after," and while accusing technology, look to it for an essential pick-me-up to get us through the future. Ironically, technology just may come through again. If so, the big question will be whether or not the human race has learned from its twentieth century experiences. Has it learned to use its technology gratefully rather than greedily? Has it learned to temper the appetites of today out of consideration for the needs of tomorrow? "The fault, dear Brutus, is not in our stars, but in ourselves," warned Cassius in *Julius Caesar*. Medieval scholar Jean Gimpel wrote that "one of the great tragedies in history is that a declining society, hoping to live more peacefully than in the past, tends to regret its technology."[9]

So we'll see what happens. We can assume that technology will shrug off criticism and go about its business of developing solutions and methods that can erase our problems and enrich our lives. Ironically, if misused, those same solutions and methods can bloat us with gluttony. But wouldn't it, in that case, be silly to blame the chef? The fault, dear Brutus.

Meanwhile, technology is acknowledging receipt of its permanent invitation to help us catch up with tomorrow in reasonably comfortable style by advancing on the multitude of energy fronts described in this book.

Professor Terry Kammash at the University of Michigan informed this writer that the prospects for fusion energy have now advanced past the theoretical stages to those of nuclear engineering. Plasma physicists have determined principles and indicated "what": the controlled fusion of atomic nuclei for dividends of energy. Nuclear engineers are currently determining "how."

Broader and more efficient applications of solar energy are also awaiting the development of new, more flexible and efficient methods. These needs too report to the province of technology and wait at the border to have their visas stamped.

Obtaining maximum benefit as quickly as possible from other major energy sources is at this point chiefly a function of "how" rather than "what."

The analogy of the mountain serves. In a sense basic science often takes men to the foot of the mountain where the shining summit (*e.g.*, fusion energy) can easily be seen.. Reaching the mountain is vitally important, but there is another task, equally important: climbing it.

If the sides of the mountain are steep, treacherous and difficult (*e.g.,* the critical requirements of a fusion reactor), nevertheless they must be safely overcome if the summit is to be reached. Providing the tools and translating principles into specifics comprise a special invitation to technology.

In fusion, the mountain has been reached and the summit is assumed to exist even if it is not yet in view. According to Professor Kammash and other experts, the mountain is in the slow but steady process of being methodically conquered.

In the case of solar energy, a sea analogy fits and calls us down from the mountain to a vast ocean of energy. Standing on the shore, we can see endless energy stretching to the horizons and beyond. The technological challenge is to concentrate portions of that energy ocean and to use them efficiently. Many scientists and engineers are diligently applying themselves to the complex and pervasive question of "how?" They may not find perfect answers, but they will find answers. It will be up to men to make the most of them, and the best of them.

ERA OF TRANSITION

Changes are taking place—inevitable changes—in the long familiar pattern of energy consumption. Certain brands of energy are being depleted. Others are being sought and tried out as replacements. All of this is to be expected. It is a coherent and predictable sequence within the nature of things. Energy phases are a recurring thread woven through human history, giving the tapestry of man's collective record a common theme. Transitions from one energy phase to another are critical events in the human experience. They were survived in earlier eras. Now we are faced with the transition again, and how we respond now and act will give shape and dimension to the future. A reading of the past gives modest nourishment for cautious optimism according to writer Lawrence Rao.

> Each century of human civilization has had its special fuel sources. Cave men had wood. The early civilized centuries relied on wood and eventually coal. The twentieth century predominantly uses oil, while the eighteenth and nineteenth centuries used coal, wood, gas, and oil. Now the oil-gas age is nearing its end. Through necessity it will die a slow, painful death because we don't yet have the technology to make use of solar and nuclear power adequately. But I have great faith in human ingenuity. There is no doubt in my mind that a change to a new and sufficient fuel source will happen. Someone or some nation will figure something out, if only to start another war. Progress, after its fashion, continues.[10]

The technologies of new energy methods have not been perfected, at least not to the extent needed for future energy on a leviathan scale. In the time to come, the same as in time past, whatever energy is available will be used, and then more requested. An active spirit of collaboration is likely to be needed among both the suppliers and the consumers of energy. Utilization will have to be tempered, especially during the next transitional decades, to achieve and maintain a balance with energy availability. Intelligent patience will be asked of all men as technology is sought to harness the wind, the tides, the heat of oceans, and the remarkable tumult of activity in the hearts of atoms.

HAZARDS AND HOPES

One vital caution should be posted: Blind trust in the agility of technology to do its work in time and turn hopeful prospects into seasoned energy performers must not tempt men to become complacent about their energy future.

The past provides adequate warning that complacency is dangerous. In that direction energy malnutrition or starvation are as likely to be found as plenty. Whatever promise is offered by fusion and solar energy, it will be forthcoming only through effort, not through a complacent holding back. One critical form of holding back can be insufficient financing. Solar energy has been available and wasted during most of man's history, and the reason in part has been complacency, taking for granted, holding back.

Man's energy future calls not for complacency, nor optimism, nor pessimism. None of the three has ever solved a practical problem. What is called for is resolution. Resolution to solve problems and to cooperate in achieving them. Resolution not to settle for easy compromises that fall short of the goal. Fusion energy is never going to respond to anything less than all the proper questions totally solved.

Resolution will also be needed not to accept environmental assaults as a reasonable price to pay for sufficient energy. With resolution man's energy future can be *made* to prove both sufficient and benign.

In a 1974 brochure the U.S. Environmental Protection Agency accused the citizens of the United States, and by extension people in the rest of the industrialized world, of being on a "quick energy binge." The price for such a binge is rapid depletion of resources and presumably a chilly morning after. When the EPA holds a mirror up to the present, some of the wrinkles become apparent:

> In the past, Americans have been successful in developing new sources of energy when old sources were no longer capable of meeting demand. We have made the transition from wood and whale oil, to coal, to oil, to gas. But we seem to be unaware of the enormous time that has been required to get new technologies working. It behooves us to reflect that it has always taken from 40 to 60 years to get a new source of energy to the point where it could supply 10 percent of the national energy needs. Our highly industrialized energy consuming economy generates mind-boggling quantities of toxic waste products which are poured daily into the air and water and on the land, creating definite health hazards for all of our citizens . . . . If we don't change we are in real danger of running out of everything while still expecting substitutes for soon-to-be depleted resources to show up, as they did many times before.[11]

This is a valid warning against complacency and waste. It is also a large sign on the highway to the future stressing the importance of not waiting too long. "In delay there lies no plenty" advised that famous energy expert, William Shakespeare.

NOW AND FUTURE CHOICE

Sometimes because of fear, men run from the facts in order not to be dominated by unpleasant truth. Also because of fear, men may do nothing to avoid the admission that there is something to do. In *Cry, The Beloved Country,* Alan Paton writes of Dawn's coming: "The great valley of the Umzimkulu is still in darkness, but the light will come there. Ndotsheni is still in darkness, but the light will come there also. For it is the dawn that has come, as it has come for a thousand centuries, never failing. But when the dawn will come, of our emancipation, from the fear of bondage and the bondage of fear, why, that is a secret."[12]

The coming of light, the resurrection of energy in new forms, is ahead of us in the future. It will come sooner and more smoothly if we don't panic, if we leave fear behind in the darkness, and do the waiting jobs. The timetable for commercial fusion is put at 50 years, but with fiercer enterprise, mountains are sometimes conquered more quickly than initial planners dared predict.

Solar energy could be established as a major future energy quickly, but not through delay, not without effort.

We are not energy poor on this planet. We have uniquely superior options in solar energy, fusion, *et al.* Yet we may become poor indeed if we are lacking in resolution, and if we are weak in will. Which way will man's future energies go, toward starvation or toward plenty?

It is up to all of us. The decision is being made now, and the suspense mounts as we wait to learn what we've chosen.

ASPECTS OF THE FUTURE—REFERENCES

1. Lovins, Amory. "Resilience in Energy Strategy," *The New York Times,* July 24, 1977, p. E17.
2. Watkins, Bruce O. and Roy Meador. *Technology and Human Values* (Ann Arbor, Michigan: Ann Arbor Science Publishers, 1977), p. 48.
3. Sullivan, J. W. N. *The Limitations of Science.* (New York: The Viking Press, 1933).
4. Wilhelm, John L. "Solar Energy, the Ultimate Powerhouse," *National Geographic,* Vol. 149, No. 3, March 1976, p. 383.
5. Hayes, Dennis. "Conservation as a Major Energy Source," *The New York Times,* March 21, 1976.
6. Carter, Luther J. "Energy Policy: Independence by 1985 May Be Unreachable Without Btu Tax," *Science,* Vol. 191, No. 4227, February 13, 1976, p. 546.
7. The National Energy Plan, Executive Office of the President, Energy Policy and Planning, April 29, 1977, pp. 35-47.
8. Marlin, Ted. "An Introduction to Recycling," *Doing More With Less,* No. 2, February 1977, p. 5.
9. Gimpel, Jean. *The Medieval Machine* (New York: Holt, Rinehart and Winston, 1976), p. 228.

10. Rao, Lawrence. Quoted in *Perspectives on the Energy Crisis,* Vol. 1, Howard Gordon and Roy Meador, Eds. (Ann Arbor, Michigan: Ann Arbor Science Publishers, 1977), pp. 5-6.

11. U.S. Environmental Protection Agency, Region V, Chicago, Illinois. "The Time Has Come," 1974.

12. Paton, Alan. *Cry, The Beloved Country* (New York: Charles Scribner's Sons, 1948), p. 273.

FUTURE ENERGY SCENARIOS

Nature is herself again. He and she walk the earth and feel the green summer between their toes. In winter they wrap their feet in skins and play in the snow. Evenings they sit in the opening of their cave and watch the naked little children throw pebbles at one another. How simple life is. How pleasant. How happy. "Just smell that air," he says. "I don't smell anything except you, dear," she replies softly. So they sit close together at the entrance to their cave watching the darkness move in. Then they listen to the grass as it slowly creeps up and over the asphalt.

"I'm off to school, mommy." "How many times must I tell you never to leave this apartment without your mask! Where's your fresh air tank? Keep just ignoring what I say and the Asphyxiation Man will get you. Then you'll be sorry you didn't listen to mommy." From the window she watched him walk through the brown mist to the subway. The boy turned to wave and in his excitement knocked the mask loose. She gasped. She could never reach him in time! But a guard noticed and quickly restored the mask. Thank goodness for the guards. She couldn't stand it if the boy, like his father, had to be put in the next "Package for the South," and dropped with the other defectives on the ice in Antarctica, so he couldn't infect others. She turned from the window to look thoughtfully at the superb conveniences of her ultramodern home.

Oil, natural gas, and coal were gone. The farewell ceremonies were held to commemorate the going. Each was given a gold watch and some nice words. What next? Nuclear was still a herd of headaches, and fusion they said was fifty years off. There were small amounts of solar energy here and there, but they were only drops in an ocean of need. "Why didn't we do something sooner?" people wondered, "instead of talk, talk, talk." They were cold in winter, hot in summer. There was no energy to do the work, so grumbling, they had to do a lot of it themselves. This forced one muscle-sore philosopher to think. And he thought of the only answer. He made his suggestion to a Congressional Committee. The Senators were shocked, but they had to agree. It was regrettable but necessary. They must reluctantly reinstate the institution of slavery. A bi-partisan group of prominent Americans was appointed by the President to designate who should be the slaves this time around.

"Mr. President, Mr. President, we just ran out of oil!" "Fine, flip the switch and turn on the sun." The switch was ready, so they flipped it, and lo the sun took over from oil. They came again and said, "We hoped it wouldn't happen. It did. We just ran out of coal." "Fine, turn on the wind, turn on the tides, and while you're at it split some atoms and bring a bowl of fusion for breakfast." "As you wish, sir." And it was done, because they had looked ahead, imitated the squirrels, and for a long time busied themselves storing up energy to be prepared when the long winter came and the messengers arrived with the bad news about oil. When the bad news arrived, they were ready, and their civilization kept ticking instead of coming to a halt.

Won't it be interesting to see which scenario we allow to win.

INDEX

188

Biogas 146-148
Biomass 144
Black hole theory (sun core) 44
Blackout, New York City 3,50,
169-170
Blake, William 3
Bohr, Niels 13
Breeder reactors 114-120
CANDU reactor 115,118
Clinch River reactor 115,119,
123
electricity from initially 111
liquid sodium coolant 121
operation 121
Phenix reactor 115,121
Superphenix 115
U.S. deemphasis 108,119,123
Brown, Governor Jerry 175
Buchanan, N.Y., site of nuclear
reactor 117
Byrd, Richard E. 122-123
Byron, George Gordon 141,142

Cadmium, liquid, nuclear reaction
control 124,126
Cambridge University 44
CANDU breeder reactor (Canada)
115,118
Carbon cycle, stellar energy source
43
Carbon-14 44
Carboniferous Period 91
Carter, President Jimmy 4,123
Cassava, Brazilian energy source
145
Celestine III, Pope, windmill tax
133
Cells, solar 73-75
Chadwick, James 13
Chicago, University of 12,124
Childe Harold (Byron) 141
Christensen-Dalsgaard, J. 44

Christian Science Monitor 99
Cities, solar 78-79
Clarke, Arthur 172
Clean Air Act, 1970 93
Clinch River, Tennessee, site of
breeder reactor 115,119,123
Coal-washing plant (Drakesboro,
Kentucky) 96
Coal 91-100
flue gas desulfurization (FGD)
95
fluidized bed combustion 97
formation 91
gasification 96-97
liquefaction 97-98
Middle Ages use 92
production increase 93,99
reduction process 95
reserves 92,99,101
scrubbers 95-96
slurry pipelines 94
smoke upsets queens 92
synthetic fuels source 100
Coal technology 95-98
Collectors, solar (*See* Solar
collectors)
Colorado State University 62
Columbus 84,133
Commoner, Barry 110
Compton, Arthur 124
Conant, James 124
Concentrators, solar 55,72,79
Congressional Record 140
Conservation 68,169,173-176,183
conservation ethic 175
presidential appeal 4
Consolidated Edison Indian Point
nuclear plant (Buchanan,
N.Y.) 117
Coppi, B. 21
Corn Is Green, The (Williams) 91
"Courting of Dinah Shadd"
(Kipling) 1

192

Lithium, world supplies   22-23
Livy   162
Löf, O. G.   62
Los Alamos Scientific Laboratory
      15,65,157
Lovins, Amory   170

"Magnetic Bottle" (plasma control)
      18
Manhattan Project   13
Manure, energy source   148
Marlin, Ted   178
"Marriage of Heaven and Hell"
      (Blake)   3
Masefield, John   161
Massachusetts Institute of Tech-
      nology   9,24,146,178
   M.I.T. Tokamak   24
Massachusetts, University of   136
Maxey Flats, Kentucky, radioactive
      leakage   117
Medieval Machine, The (Gimpel)
      162
Meitner, Lise   12
Meltdown, atomic reactor threat
      113,114
Methane, hydrocarbon fuel   100
   from algae and plants   143
   from manure   148
   problem in coal mines   99
Micawber, Mr. (*David Copperfield*)
      174
Michigan, University of   15,35,
      105,180
Minimum Energy Dwelling   68
Mobil Corporation   51
Mobil Tyco Solar Energy Corpora-
      tion   75
Money crisis   1
Monsanto Company   147
Moon rotation   160
Moore, Henry (statue "Energy")
      12

Muntzing, L. Manning   125

NASA   55,61,74,143
National Academy of Sciences
      106
National Canners Association   49
National Energy Plan, U.S., 1977
      2,49,74,77,98,108,119,143,
      157,166,175
National Geographic   52
National Science Foundation, U.S.
      61,140,143,145
National Uranium Resource Evalua-
      tion Program (NURE)   106
Natural gas   99,148
   depletion   91,93
   from coal/shale   99
Natural Gas Pipeline Company
      148
Nero, Emperor   7
Neutrinos   43
Neutron bomb   33
New Mexico State University,
      windmill course   133
New Republic   84
New York City blackout   3,50,
      169-170
New York Times   5,8,52,53
Niels Bohr Institute   138
NSF/NASA Solar Energy Panel
      143
Nuclear Fission   105-128
   blessing or curse   37
   liquid cadmium control
      124,126
   safety   111,125-128
   schematic   112-113
   Scientists' Declaration   110
   scientists for and against   109
Nuclear Fuel Services plant failure
      (West Valley, N.Y.)   117
Nuclear reactor   107-120
      (*See* also Breeder reactor)

Tomlinson, H. M.   153
Toroidal pinch (plasma confine-
     ment)   19
   Stellarator toroidal machine   21
Tritium (fusion)   22-23
   radioactivity   36
   world reserves   22
TVA   96
TwoX-IIB (2X-IIB) mirror machine
     (fusion)   24

United Nuclear Corporation   110
United States Geological Survey
     (USGS)   99,155,158
Uranium   109,112,114,121,125
   dwindling supply   119
   enrichment process   108,112,
     121,123-124
   National Uranium Resource
     Evaluation Program (NURE)
     106
   reserves   106

Valdez, Alaska, oil port   4
Vocabulary, solar   88-89
Volcanoes   (*See* also Geothermal
     energy)   153-155
Vostok, Russian Antarctic base
     122

Wald, George   33,109,114

Walden (Thoreau)   81
Waste Makers, The (Packard)   177
Water, in coal production   94
Water power   163-168
   potential   164
Watson, Dr.   2,43
Watson, James   109
We Almost Lost Detroit (Fuller)
     126
Western New York Nuclear Service
     Center   35,117
West Valley, New York, site of
     radioactive waste reprocess-
     ing   35,117
White, Don   155
Wilhelm, John   75
Williams, Emlyn   91
Williams, J. Richard   46,50,53,54,
     57,67,134,139,140
Wind, energy source   132-138
   environmental effects   136
   wind generators   134-138
Windmill   132
   antiquity   132-133
   New York City application   50
   taxed by Pope Celestine III   133
   university course   133
"Windmillers"   133
Wisconsin, University of   67
Wood, Peter, solar home   64
World Meteorological Organization
     136

More Energy Books From
ANN ARBOR SCIENCE

SOLAR ENERGY: TECHNOLOGY AND APPLICATIONS (Revised Edition) — Williams

The Consumer's ELECTRIC CAR — Wakefield

1977 SOLAR ENERGY & RESEARCH DIRECTORY

EXTRACTION OF MINERALS AND ENERGY — Deju

COMBUSTION: FORMATION AND EMISSION OF TRACE SPECIES — Edwards

ENVIRONMENTAL ASPECTS OF NUCLEAR POWER — Eichholz

FUTURE ENERGY ALTERNATIVES (Revised Edition)—Meador

PERSPECTIVES ON THE ENERGY CRISIS, Vol. 1 — Gordon/Meador

PERSPECTIVES ON THE ENERGY CRISIS, Vol. 2 — Gordon/Meador

THE CONCEPT OF ENERGY — Hoffman

FUEL AND THE ENVIRONMENT — Institute of Fuel

THE UNSETTLED EARTH — Jones

FUSION REACTOR PHYSICS — Kammash

POWER GENERATION: AIR POLLUTION MONITORING AND CONTROL — Noll/Davis

EMISSIONS FROM COMBUSTION ENGINES AND THEIR CONTROL — Patterson/Henein

SOLAR DIRECTORY — Pesko

ENERGY, AGRICULTURE AND WASTE MANAGEMENT — Jewell

FUELS, MINERALS & HUMAN SURVIVAL — Reed

INTRODUCTION TO ENERGY TECHNOLOGY — Shepard

TECHNOLOGY AND HUMAN VALUES — Watkins/Meador

THERMAL PROCESSING OF MINERAL SOLID WASTE FOR RESOURCE AND ENERGY RECOVERY — Weinstein/Toro

SCIENCE AND TECHNOLOGY OF OIL SHALE — Yen